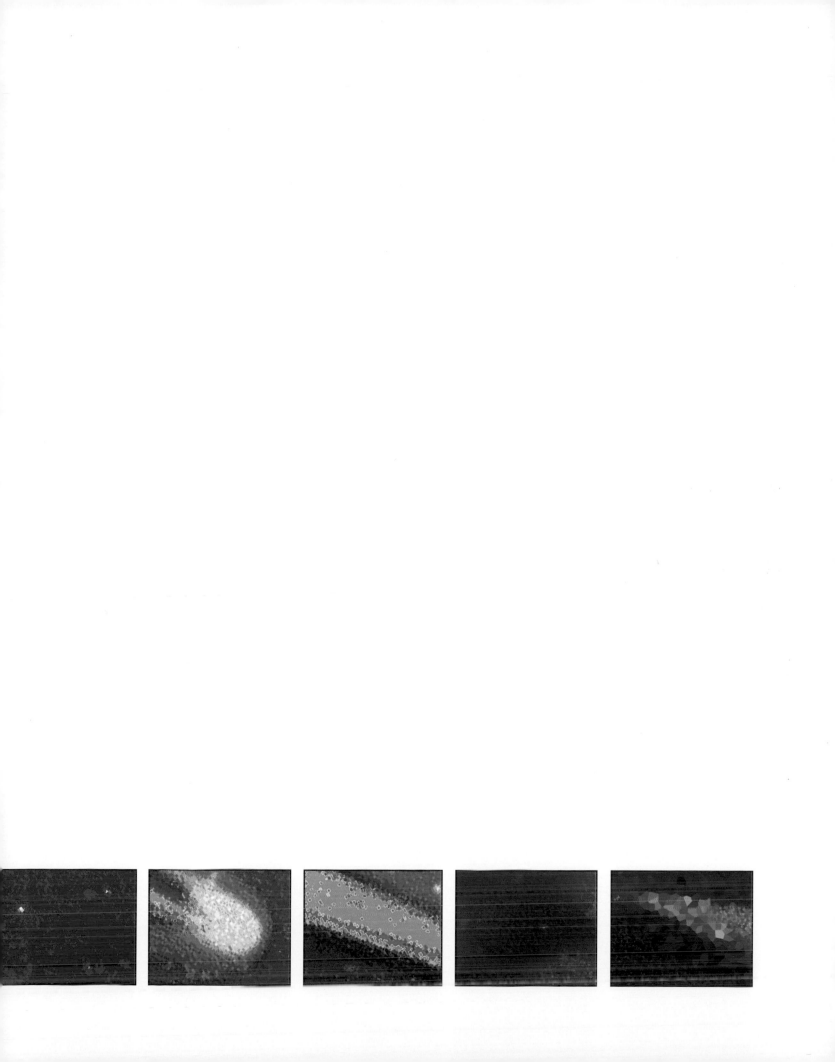

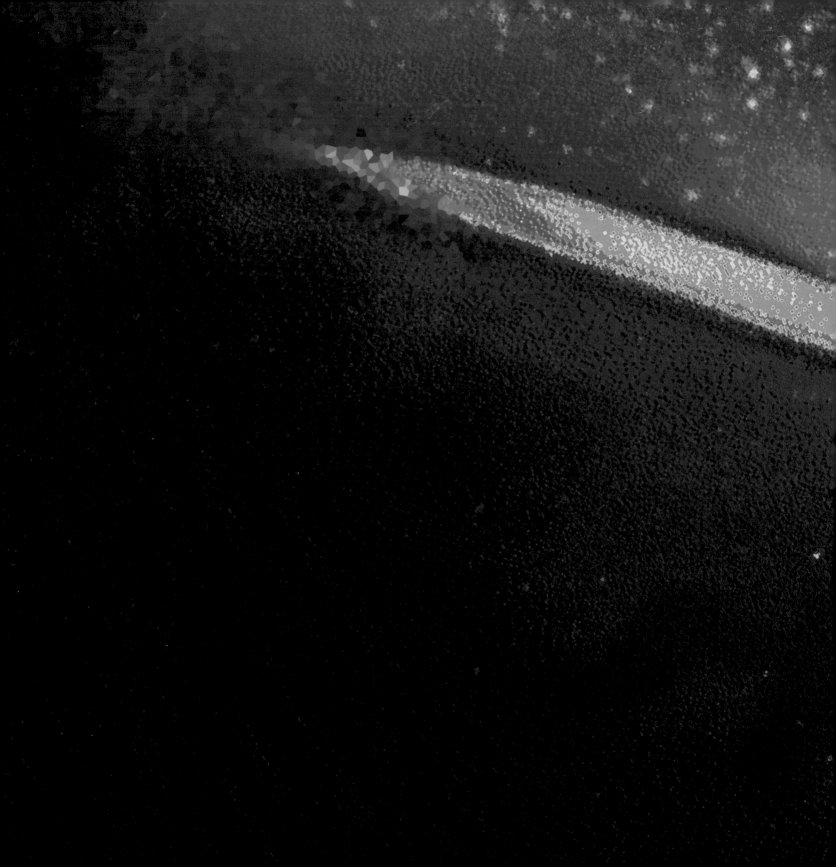

METEORITES

A JOURNEY THROUGH SPACE AND TIME

ALEX BEVAN AND JOHN DE LAETER

CONTENTS

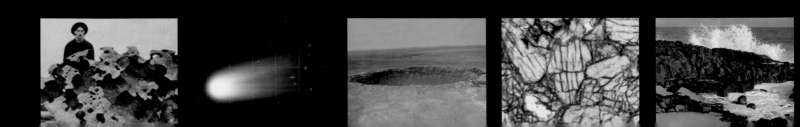

CONTENTS

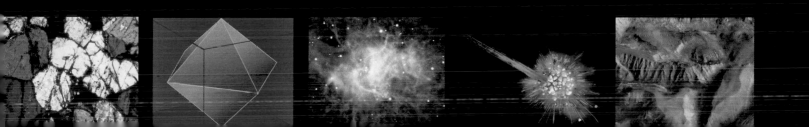

This book is dedicated to those people who have presented meteorites to museums for research and display, in the recognition that their discoveries are of
immense value to science, and represent part of our common heritage.

Published in the United States of America
by the Smithsonian Institution Press
in association with the University of New South Wales Press
University of New South Wales
Sydney 2052
Australia

Library of Congress Cataloging-in-Publication Data
Bevan, Alex.
Meteorites : a journey through space and time
/ Alex Bevan and John de Laeter.
p. cm.
Includes bibliographical references and index.
ISBN 1-58834-021-X (alk. paper)
1. Meteorites. I. Laeter, J. R. de. II. Title.
QB755.B48   2002
523.5'1—dc21   2001049551

Manufactured in China, not at government expense
09 08 07 06 05 04 03 02   5 4 3 2 1

PREFACE

This book is written for people who are not scientists.

Many people tell us that they are put off science by its difficult language: its special terms, jargon, and acronyms (words made from initials). To a scientist, all this is a necessary evil. These words often cover very precise concepts, giving a lot of meaning without going into all the explanation. This allows scientists to communicate complex ideas to each other without having to repeat all of the original, and sometimes very long, definitions.

When it comes to writing science for non-scientists, this language can be a problem. Books must either leave it out, or simplify it in some way. Inevitably, some of the lengthy explanations needed are too long, and have to be shortened. In so doing, some of the original meaning may be lost, or slightly altered. Simplifying science can lead to stating something as fact, without presenting any of the supporting evidence. This is as unsatisfactory as finding someone guilty without having a trial!

Unfortunately, the study of meteorites calls on a number of different sciences, and their scientific languages. In this book, this language is kept to a minimum. Where special terms and jargon are used, they are explained. We have tried to avoid acronyms, although we do not claim complete success. There is also a Glossary of all these terms at the end of this book.

Meteorites: A Journey through Space and Time includes much of the evidence on which our current understanding of meteorites and planetary science is based. The fourteen chapters are essentially essays, supported by boxed information and illustrations. We hope this book dispels many of the popular misconceptions about meteorites. After all, what use is knowledge if it cannot be shared?

INTRODUCTION

How many times have you looked up on a clear night to see a spark of light streak across the sky, then fade into the blackness? Popularly known as 'shooting stars', these celestial fireworks are meteors. They result from tiny pieces of natural space debris hurtling through the upper atmosphere and burning up. Tens of thousands of tons of extra-terrestrial dust settle on Earth each year only to accumulate in the oceans, or become lost on the surface of our rocky planet.

Some of the larger objects fall to Earth intact, as 'meteorites'. They are occasionally recovered by people who see them falling, or by scientists out hunting for them. Mostly they end up in museum collections, carrying the names of the places where they fell or were found. Meteorites are an endless source of fascination for everyone, and a treasure trove of information for scientists. But what is it that this debris from the Solar System tells us?

Of the planets in the Solar System, Earth is the most active, geologically speaking. Since it formed, our planet has changed constantly, rubbing out many of the clues to its early history. The cratered surface of the Moon tells of a period of heavy bombardment by huge chunks of Solar System debris more than 3800 million years ago. Yet on Earth, although some younger impact craters are known, constant geological forces have wiped the record of this early event from its surface. In that same 3800 million years, continents have arisen, seas and oceans have opened and closed, mountains have been pushed up only to be eroded down to their roots again by wind and water, and volcanoes have erupted new rock from the Earth's hot interior. Ancient rocks in continents such as Australia, Africa and North America formed as long ago as 3800 million years. Other rocks, from Western Australia, contain a legacy of mineral grains worn from pre-existing rocks that existed 4400 million years ago. But clues to the nature of the planet before this time are nowhere to be found in the rocks that now make up the Earth's surface.

What were the original materials from which the Earth was made? To get an idea of what these might have been like, we have to look at meteorites — fragments of rock (stony meteorites), metal (iron meteorites) and mixtures of rock and metal (stony-iron meteorites) — which have survived their fiery descent from space to Earth. As well as meteorites, a large amount of meteoritic dust, harder to recognise and collect, is an important supplement to our knowledge of the Solar System and beyond. The stuff that many meteorites are made of is 4555 million years old, having remained virtually unaltered since its formation. Meteorites mostly represent the debris left over after the formation of the planets and, like messengers across space and time, they carry a unique record of the earliest events in the birth of the Solar System.

A few meteorites containing water and rich in complex compounds of carbon, oxygen, nitrogen and hydrogen may represent the original materials from which our planet gained water for the oceans, gases for the atmosphere, and other essential ingredients for the evolution of life. Opening a window on the complexities of star formation, tiny diamonds and other grains found in meteorites record events that happened long before the Solar System was born.

Like detectives, planetary scientists search for clues to unravel the events of the past. They interpret the evidence and, so far as they can, provide the answers, or construct theories based on all the facts that can be accumulated by observation, measurement and experiment. Providing us with an understanding of our most distant past, meteorites are of great scientific interest, and their study (now called meteoritics) has played an important role in advances in physics, chemistry, geology and astronomy.

So what are meteorites made of? Where do they come from? How and when did these materials form and, importantly, what do they tell us of the nature of the infant Solar System? This book contains many examples from around the world, explores the evidence from meteorites, and examines some remarkable discoveries made over two centuries of scientific enquiry.

... the existence of such a substance as iron had long

been known to the Egyptians as a natural product.

They obtained it from meteorites, called it bia,

and practically the only use they had made

of it was for magical or ornamental purposes.

GA WAINWRIGHT, 1932

ANCIENT BELIEFS

The Old and New Worlds

Late in 1911, at El Gerzeh near the banks of the Nile about 70 km south of Cairo, British archaeologist Gerald Wainwright discovered an ancient Egyptian cemetery. Sifting through the contents of an undisturbed grave with his friend Joscelyn Bushe-Fox, they found the remains of a young adult surrounded by pottery. The skeleton was lying north-south, on its left side. The head had been removed and placed upright at the southern end, facing the west. Wainwright recognised the style of burial and grave contents as dating from the middle of the 'pre-dynastic' period of early Egypt, around 5500 years ago. The bones were cracked and crumbling, but around the neck and waist were two strings of beads made from gold, carnelian, agate — and iron. Gold and other minerals used for jewellery are not unusual in royal graves of this age, but the iron beads were a startling discovery.

These iron beads, along with more found in another grave, came from a time when the Egyptians were still using stone tools. This was long before they had mastered the art of extracting iron from minerals, or developed their distinctive hieroglyphic writing. Now deeply rusted, the beads had been made by beating pieces of metal into thin plates, then bending them into tubes. Wainwright realised that these early Egyptians' experiments with iron must have influenced them greatly. Analysis showed that the beads contained a substantial amount of nickel,

a characteristic of meteoritic iron. What Wainwright and Bushe-Fox had found was one of the earliest known examples of human contact with meteorites.

The Egyptians' use of metal hacked from meteorites is not unique. Later archaeological finds confirm that many other civilisations throughout the world have done the same. The oldest known examples, dating back more than 6000 years, are meteoritic iron balls found in a grave at Tepe Sialk in Iran.

Following the discovery at El Gerzeh, metal fragments found in excavations at the royal cemetery of the Sumerian city of Ur of the Chaldees (now in Iraq) and dating back more than 4500 years were also found to be meteoritic. Some of the oldest civilisations observed meteorite falls, collected the debris and wrote accounts of the events. The oldest known record of a meteorite fall comes from around 4000 years ago in Phrygia (now part of Turkey). According to the Roman historian Titius Livius, the celebrated meteorite at Phrygia was later transported in royal procession to Rome where it was worshipped for another 500 years.

From their early writings, it is clear that the Egyptians, Greeks, Romans and Chinese understood that meteorites fell from the sky. Modern studies of ancient texts reveal that the Hittites, an ancient people who lived north of present day Syria, recognised an extra-terrestrial origin for meteorites at least 3200 years ago. Their word for iron, *kù-an*, may be one of the earliest names for meteoritic iron.

The word *meteorite* comes from the ancient Greek *meteoros*, meaning 'originating in the

Left: Jewellery containing meteoritic iron beads was found around the neck and waist of a skeleton in a grave uncovered in 1911 by George Wainwright at Gerzeh in Egypt.

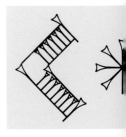

The Sumerian symbol (sumerogram) pronounced kù-an, possibly one of the earliest terms for meteoritic iron, appeared towards the end of the third millennium BC.

atmosphere'. An Egyptian hieroglyph, possibly pronounced *bith*, means 'heavenly iron'. The importance of meteorites in ancient Egyptian culture is embodied in the 'Opening of the Mouth' ceremony — a solemn part of burial rituals. Special tools for this practice included *netjeri*-blades, made principally from meteoritic iron. With gaping mouths, the dead were symbolically brought to life and a pathway opened to heaven for the *ka*, or soul.

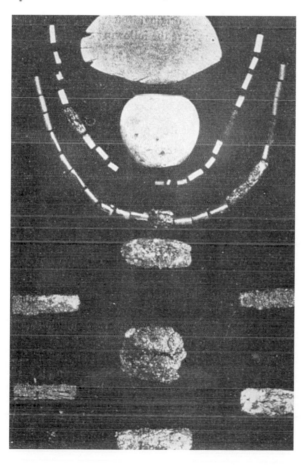

Ancient texts, together with coins struck to celebrate the falls of meteorites in antiquity, suggest that some civilisations eventually gave up working meteorites for tools and jewellery. Instead they preserved them for worship. Exactly when meteorite worship began is unknown, but the practice became widespread among civilisations bordering the Aegean and Mediterranean more than 2000 years ago, so much so that meteorite cults developed and flourished for many centuries.

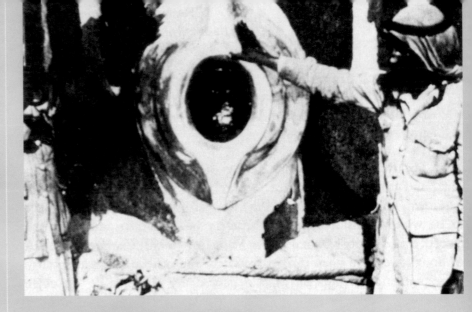

The mystery of the Black Stone of the Ka'bah

The holy stone of the Ka'bah, treasured by Muslims for at least 1500 years, is the most venerated rock on Earth, and may have a meteoritic connection. Today the stone lies in the Great Mosque in Mecca, at the eastern corner of the holy building called the Ka'bah.

According to Muslims, the origin of the Black Stone can be traced back to Adam. Legends tell that Adam built the first Ka'bah. A white stone, later to become black through the sins of mankind, served Adam as a chair. The stone is believed to have remained in the Ka'bah until the building was destroyed by the biblical flood, and it was later returned to Abraham who rebuilt the Ka'bah. Other legends say that Abraham built the first shrine. Written accounts of Arab historians record that the Black Stone has suffered a remarkable history — having been burned, stolen and smashed. In 1050 AD, an Egyptian caliph sent a man to destroy the stone, but the fragments survived.

During the nineteenth and twentieth centuries, accounts were published about the origin of the Black Stone by European travellers to Arabia, a few of whom, disguised as pilgrims, had seen it first hand. Conflicting reports identify the Black Stone as a stony meteorite, basalt lava, an agate and, more recently, fusion glass from the meteorite impact craters at Wabar in Saudi Arabia. Originally a single stone, today the Black Stone appears to be made up of eight pieces cemented together and surrounded by a silver frame. Writing in 1875, an Arab historian, al-Kurdi, reported that 50 years earlier it had consisted of 15 fragments. The missing fragments may now be hidden behind the silver frame, or they may be lost.

Veneration of the Black Stone, and the resolute efforts of Muslims to preserve it, has prevented scientific examination. Interest heightened in 1938 when Mohammed Khan, a Muslim scholar, re-examined suggestions that the Black Stone was a stony meteorite. Since then, several researchers have tried in vain to verify the claim and have met a veil of secrecy. An engraving, photographs and first-hand descriptions are the only available evidence of what the Black Stone might be.

However, the Black Stone's attributes offer some clues. The fact that it can be broken easily shows that

The Black Stone of the Ka'bah in Mecca mounted in a silver frame.

A group of craters was formed by the explosive impact of an iron meteorite around 6000 years ago at Wabar in Saudi Arabia. The largest crater measures 100 m across. This smaller 11-m crater was found only recently.

it is not an iron meteorite. Suggestions that it must be a hard rock stem from its appearance, and a mirror-like polish (that may have been caused by the rubbing of generations of pilgrims' hands). Significantly, according to the testimony of an Arab, the Black Stone is said to float in water. This made it possible to recognise the Black Stone in 951 AD when it was returned in two pieces from an earlier theft. Basalt lava, agates and stony meteorites do not float, although a sizeable lump of volcanic pumice would.

While the meteorite theory is discounted, the colour of the stone has been difficult to establish. Most descriptions say that it is brownish-black or deep reddish-brown. The change in colour from the original white could be explained by the thousands of years of handling. Some descriptions note that the

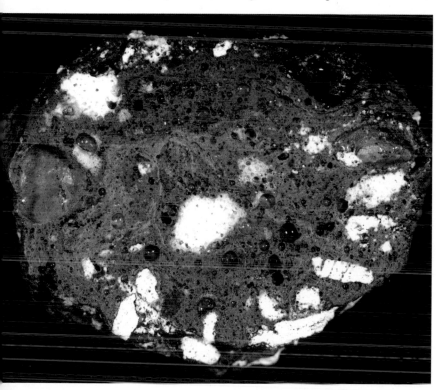

rothy glass 'bombs'
were made when sand
melted during the
impact and excavation
f the Wabar meteorite
raters. A similar
omb, coated with
hiny black iron oxide,
us been suggested
s the source of the
lack Stone of the
a'bah.

stone has white or yellow spots that appear to be scattered crystals. One account records that the interior is as 'white as milk'. None of these features would rule out volcanic, or some other natural glass.

Although it cannot be a meteorite, the Black Stone is regarded by Muslims as 'an object from heaven', and there may be a link. The riddle of the Black Stone may have been solved in 1980 when Elsebeth Thomsen of the University of Copenhagen suggested that it is glass from a meteorite impact crater. Approximately 1100 km east of Mecca, at Wabar in Saudi Arabia, is a group of craters that were

formed by the explosive impact of an iron meteorite around 6000 years ago. The craters were discovered in 1932 in a region known as Rub'al Khali, meaning 'the Empty Quarter'. Iron meteorites have been found around the craters that are situated in a pure, whitish sandstone. A feature of the craters is the presence of blocks or 'bombs' of silica glass, partly impregnated with tiny beads of once-molten nickel-iron alloy that formed during the blast that destroyed most of the projectile, melted the country rocks and excavated the craters.

Some blocks of pumice-like Wabar glass are made of frothy white interiors with black glassy shells. The white material is fragile, but the iron-rich black glass is quite hard. Other bombs are mixtures of large fragments of partially fused sandstone, minerals and dark glass. Importantly, the glass contains gas bubbles that would allow it to float on water. The reported age of the Wabar craters is well within the time of human occupation of the Middle East, and it is possible that the impact was witnessed. Thomsen suggests that the glass could have been transported to Mecca along a caravan route from Oman that may have passed close to the craters.

In ancient Arabic poetry Wabar (or Ubar) is the site of a legendary city destroyed by fire from heaven because of the wickedness of its king. Is the Holy Stone of the Ka'bah a piece of Wabar impact glass? Recent estimates of the age of the craters suggest that they may have formed less than 450 years ago, ruling it out as the source of the stone. Proof hangs on an examination of the stone and an accurate age for the craters. Futile attempts to obtain a sample of the stone for analysis led one scientist, Lincoln La Paz to write:

> My disappointing experience in this venture leads me to believe that procurement of even scientifically usable photographs of the Black Stone (let alone acquisition for chemical and microscopic study of one of the many small fragments into which it has been broken) will long remain an unattainable goal of the meteoriticist — be he a Western unbeliever or an Islamite.

The statement remains true 30 years after he wrote it.

Meteorite use and abuse

More than thirty venerated meteorites are known across the Old and New Worlds: in Argentina, Mexico, Africa, India, the Middle East, Estonia, Russia, Japan, the USA and Australia. Many examples come from North America, where a variety of

The cavernous mass of the Willamette iron meteorite was venerated by generations of North American Indians. It is now kept in the Smithsonian Institution's National Museum of Natural History in Washington, but even today members of the Clackamas tribe visit to pay homage.

indigenous tribes paid tribute to large and, for them, immovable masses of meteoritic iron. The most famous is the 14.1-ton mass of the Willamette iron meteorite that was considered powerful medicine by the Clackamas Indians of Oregon. Indians washed their faces in the water that collected in the basins of this large cavernous mass, and young warriors dipped their arrows in the water before going off to war with neighbouring tribes. Another is a 2-ton meteorite found in Arizona. Thought to have been a sacred monument of the Navajo tribe, it now carries their name in the Field Museum of Natural History in Chicago.

Among the ancient Hopewell Indians of North America there was an early trade in meteorites. Evidence lies in beads and artifacts made from a stony-iron meteorite found early in the nineteenth century in the great Hopewell burial mounds in Ohio. Several researchers have since shown that the source of the material was the Brenham meteorite, found in south central Kansas in 1882, some 1400 km west of the mounds. Precisely when this trade took place is unknown, but more than 2000 years ago the late stone-age Hopewell Indians established a trade route along the Arkansas and Missouri Rivers carrying with them, along with many other things, a precious cargo of meteoritic metal. Confirming the Hopewell Indians' interest in meteorites, 29 beads of an iron meteorite were found in 1945 during the excavation of other burial grounds at Havana in Illinois.

Before humans mastered iron technology, the chance discovery of meteoritic iron presented a great prize. Many finds were beaten and worked into

implements and weapons. Several Chinese artifacts, including two early weapons, contain meteoritic iron. A broad axe with a spike made of bronze and a deeply rusted meteoritic iron blade, along with a dagger axe pointed with meteoritic iron, date from the early Chou Dynasty, about 3000 years ago. These weapons may have come from the tomb of Prince of Wei in Honan Province. Significantly, they were manufactured about 400 years before the general appearance of cast iron metallurgy in China.

Because meteoritic iron varies in composition, and sometimes contains abundant impurities, it is notoriously difficult to forge. Nevertheless, there are many ancient and modern examples of ceremonial weapons made from meteoritic iron, the metal having been saved for the swordsmith's finest work. The Prambanan meteorite from Indonesia, of which only

A drawing by the interpreter Zachaeus shows the HMS Isabella and Alexander at ice-anchor east of Cape York, while Captains Ross and Sabine trade jewellery and clothing for narwhale tusks and meteoritic iron-tipped implements with the Inuits

a small portion remains, is one example. Following their fall before 1797, two masses of the Prambanan iron were kept in the sultan's palace in Soerakarta, Java, providing metal for weapons and tools. By heating the meteorites, small ingots of metal were removed by arduous chiselling. Among the forged items are superbly fashioned *kris* daggers.

Perhaps no other meteorite has been so closely entwined with the daily lives of a group of nomadic people as the Cape York irons used by the Inuits of Melville Bay in north-west Greenland. At least eight masses of meteoritic iron, totalling more than 59 tons were found among the glaciers and rocky outcrops of this remote and inhospitable part of the world. This great shower of iron meteorites probably fell long before the arrival of people in the area around 1000 years ago. Without wood, the Inuits had learned to make hunting weapons out of the bones

of walrus and reindeer, and also to exploit their precious source of meteoritic iron to tip harpoons and edge their knives. Later, in 1818, bone-handled knives edged with iron were traded with Captain John Ross when his icebound ships were visited by this previously unknown group of people.

Ross discovered that there were several masses of iron in the general vicinity, from which flakes of metal had been removed. Seventy-six years later the Inuits revealed to Robert E Peary, the famous American explorer, the locations of three of the meteorites. The Inuits called the masses *Ahnighito* or Tent (31 tons), Woman (3 tons) and Dog (0.4 tons). Dog is a large fragment broken from the 'Woman' mass over the centuries that the Inuits worked the irons.

An estimated 10 000 rounded blocks and fragments of discarded hammer stones made of basalt

Strips of meteoritic iron, possibly from the Prambanan iron meteorite that fell some time before 1797 in Java, Indonesia, were infolded and forged with man-made iron to make *kris* daggers.

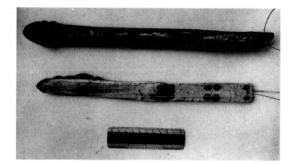

Above: The Inuit knives presented to Captain John Ross in Greenland in 1818 were made of bone, edged with meteoritic iron from the Cape York shower.

The 20-ton 'Agpalilik' from the Cape York iron meteorite in Greenland. Several other masses were exploited by generations of Inuits.

have accumulated around Woman forming a pile 8–10 m in diameter. For tens of generations, Inuits visited the site to replenish their stocks of metal. Because the local rocks were unsuitable for the arduous work of pounding the cosmic steel, each sledge party dragged new hammer-stones of basalt from a point 160 km away. By Peary's time, the meteorites were no longer being used for implements. After a great physical and financial struggle, Peary eventually succeeded in transporting the three masses to New York where they sit today in the American Museum of Natural History.

Additional finds of the Cape York shower were made in 1913 when a mass, called Savik I (3.4 tons), was located on a promontory 10 km east of the Woman and Dog masses and later taken to Copenhagen. In 1961, a small mass of 7.8 kg, Savik II, was discovered by an Inuit on a hunting trip. Later in 1963, a new 20-ton mass, Agpalilik, was found partially covered by large boulders, by one of the world's leading authorities on iron meteorites, Vagn Buchwald of the Institute of Metallurgy in Copenhagen. This mass, also taken to Copenhagen, had not been worked and was apparently unknown to the Inuits.

Perhaps the most remarkable find of all had already been made in 1914, but not recognised for what it was. A small mass of hammered meteoritic iron weighing 1.6 kg was found near an ancient igloo site on the Knud Peninsula in Ellesmere Land, Canada. Buchwald later examined what became known as the Akpohon mass, showing it to be a fragment transported some 600 km to the west of Cape York.

The practicality of the Inuits aside, early peoples' reactions to meteorites varied considerably, although there is a recurring theme of awe, fear and reverence. In Argentina, the 115-kg Caperr iron meteorite was considered taboo by Patagonian Indians, and the 1.5-ton Casas Grandes iron from Chihuahua, Mexico, was found swathed like a mummy in an Indian tomb. Another chamber of the tomb contained similarly wrapped human remains. In Africa, the colossal Mbosi meteorite, estimated to weigh around 16 tons, is now a national monument at the place where it was found, but was reported by early European travellers to have struck fear into indigenous Tanzanians. Although the meteorite probably fell in antiquity, when it was discovered in 1930 there were no signs of its exploitation for metal.

Surprisingly, for one of the world's oldest surviving civilisations, there is scant evidence of interaction with meteorites by Australian Aborigines in antiquity. Possible reasons for this are the unsuitability of most stony meteorites for tool making and the inability, or lack of need, of early Aborigines to work metal. But Aboriginal use of meteorites was other than practical.

Among the earliest recoveries of meteorites in Australia were two large masses of iron weighing 3.5 and 1.5 tons found near Cranbourne in Victoria in 1854. During the period from 1854 to 1928, eight additional masses of the same meteorite shower were found, bringing the total weight recovered to more than 10 tons. Early European colonists recounted a time when Aborigines used to dance around the largest meteorite, 'beating their stone tomahawks against it, and apparently much pleased with the metallic sounds thus produced'. It is very likely that this practice had been handed down from generation to generation, while the deeply rusted state of the meteorite suggests that it fell in pre-historic times. The largest mass of the Cranbourne meteorite was removed in 1863 to the Natural History Museum in London.

Early accounts tell us that many ancient peoples accepted meteorites as heavenly bodies that were certainly not of this Earth. As with many other aspects of knowledge, from an early, enlightened understanding by the ancients of the nature of

An imaginative woodcut depicts the fall of a meteorite.

'Of the thunder-stone that fell in the year 1492 in front of Ensisheim.'

meteorites, the Dark Ages brought to Western culture superstition and ignorance.

Records of meteorite falls between the times of ancient Egypt and the Middle Ages are few. Venerated for over 1000 years, the oldest observed meteorite fall is enshrined in a Shinto temple at Nogata-shi in Japan. Records show that this stone fell in the temple grounds on 19 May, 861 AD.

The oldest mediaeval meteorite fall for which material is still generally available to science fell at Ensisheim in Alsace (now a part of France) on 16 November, 1492. At around 11.30 am, after loud thunder-like noises, a stone weighing 127 kg was seen to fall. The German king, Maximilian, viewed it as divine intervention and a premonition of victory against his enemies the Turks. Maximilian commanded that the meteorite should be preserved in the church at Ensisheim. Today a large surviving remnant is on display at the town's Palais de Régence. Fragments taken from the Ensisheim meteorite have found their way into museum collections around the world.

In the year after the fall at Ensisheim, Maximilian was made Emperor of the Holy Roman

Empire. Remarkably, nineteen years later he came close to another, more spectacular meteorite fall. During the summer of 1511, the French and their allies were at war with the forces of Pope Julius II, and they had occupied Genoa, Ferrara, Milan and a portion of Lombardy, in Italy. On 4 September 1511, after the appearance of a great fireball, a huge shower of meteorites fell near the small town of Crema in Lombardy, also held by the invading French. Their adversary was none other than Maximilian. Following in the wake of the fireball, a luminous trail apparently remained visible in the night sky for more than two hours. None of the Crema meteorites are preserved, even though one may have weighed more than 50 kg. But the fireball is depicted in Raphael's celebrated painting the *Madonna di Foligno*.

The birth of a science

Although many meteorites fell in Europe and Russia during the seventeenth and eighteenth centuries, not all were preserved. Despite a renaissance in scientific enquiry, few scholars would dare to suggest that meteorites came from beyond our planet. Those who did were considered either fools or heretics, and were derided and scorned. The problem was that no natural historian had actually observed a meteorite fall. In typically conservative fashion, the scientific

Some historic meteorites

Name	Locality	Classification
Albareto	Emilia-Romagna, Italy	stone, L4
Benares	Uttar Pradesh, India	stone, LL4
Bjelaja Zerkov	Kiev Province, Ukraine	stone, H6
Brenham	Kansas, USA (Hopewell Mounds)	PAL
Canyon Diablo	Arizona, USA	IAB
Cape York	West Greenland	IIIAB
Caperr	Chubut, Argentina	IIIAB
Casas Grandes	Chihuahua, Mexico	IIIAB
Chilkoot	Alaska, USA	IIIAB
Cranbourne	Victoria, Australia	IIICD
Crema	Lombardy, Italy	stone?
Ensisheim	Alsace, France	stone, LL6
Gerzeh	Egypt	iron
Glorieta Mountain	New Mexico, USA	PAL
Havana	Illinois, USA	IIICD
Huizopa	Chihuahua, Mexico	IVA
Iron Creek	Alberta, Canada	IIIAB
Kaalijarv	Saaremaa, Estonia	IAB
L'Aigle	Orne, France	stone, L6
Livingston	Montana, USA	IIIAB
Mbosi	Tanzania	iron
Mesa Verde	Colorado, USA	IAB
Morito	Chihuahua, Mexico	IIIAB
Navajo	Arizona, USA	IIAB
Nedagolla	Andhra Pradesh, India	iron
Nogata	Kyushu, Japan	stone, L6
Oktibbeha	Mississippi, USA	IAB
Prambanan	Java, Indonesia	iron
Red River	Texas, USA	IIIAB
Salles	Rhone, France	stone, H6
Siena	Tuscany, Italy	stone, LL5
Tamentit	Tuat, Algeria	IIIAB
Tepe Sialk	Iran	iron
Thunda	Queensland, Australia	IIIAB
Ur	Chaldea, Iraq	iron
Uwet	Cross, Nigeria	IIAB
Wabar	Rub'al Khali, Saudi Arabia	IIIAB
Weston	Connecticut, USA	stone, H4
Wichita County	Texas, USA	IIICD
Willamette	Oregon, USA	IIIAB
Wold Cottage	Yorkshire, England	stone, L6

establishment was reluctant to accept the testimony of ordinary people, many of whom had given detailed and accurate accounts of falls.

Popular explanations for meteorites included their formation in the atmosphere by lightning, concretions of volcanic dust, even acts of the devil. Since most meteorite falls are accompanied by thunder-like sounds, and the light from a fireball could be mistaken for lightning, the first two explanations — although wrong — at least had some basis in logic. Other explanations of the mystery of meteorites were clearly religiously driven.

During the second half of the eighteenth century, perhaps to maintain their reputations, some scientists simply denied the existence of meteorites. This denial hindered the development of an understanding of meteorites for many years. Worse still, the dismissal of meteorites as a serious subject of research resulted in the removal of some meteorites from collections across Europe as 'rubbish'.

Two notable exceptions to the conspiracy of silence were Abbé Domenico Troili and Ernst Florens Friedrich Chladni. Troili described a 2-kg stony meteorite that fell near Albareto in the Romagna area of Italy in July 1766. Although he did not doubt that it had fallen from the sky, Troili thought it had originated in the atmosphere. Chladni was a reputable physicist, and in 1794 he published a book about a stony-iron meteorite that had been found in Russia and four iron meteorites that were then in collections. He also documented many reports of the falls of stony meteorites. Although Chladni suggested that both stones and irons were extra-terrestrial, he was unable to prove a direct link between the two. He speculated, however, that irons had come from the break-up of much larger masses. Remarkably, on the basis of very little evidence, Chladni had come to the correct conclusion — more than a century before it became generally accepted.

Gradually, from around 1790, a theory for the origin of meteorites from beyond the Earth moved from the realms of fiction to respectably established scientific fact. Adding fuel to the debate, several meteorite falls were witnessed and recovered during this period. They included falls of stones at Siena in Italy in 1794, Wold Cottage in England in 1795, Bjelaja Zerkov in Russia in 1796, Salles in France and Benares in India, both in 1798.

Exactly who was responsible for the birth of the science of meteoritics is disputed. Most historians of science credit Chladni, others acknowledge the contribution of French physicist Jean-Baptiste Biot.

Ernst Florens Friedrich Chladni was one of the fathers of the modern science of meteoritics.

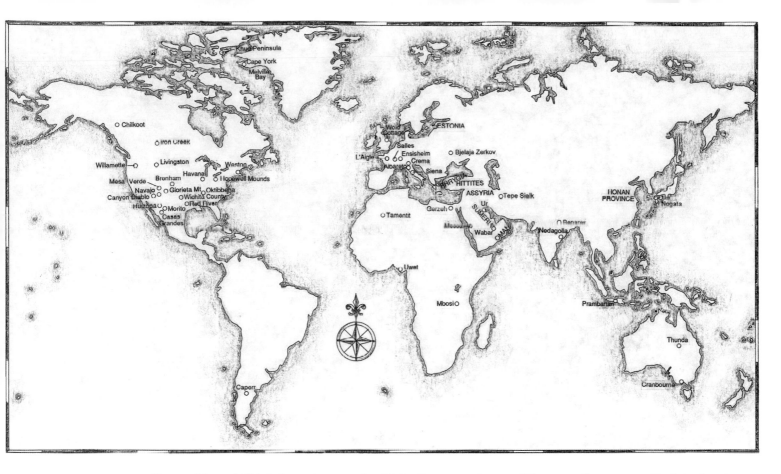

Recognition of Chladni's work came after 26 April 1803, when a large shower of several thousand stony meteorites fell near the town of L'Aigle in northern France. Importantly, this fall had been witnessed by a number of officials. Biot was hastily sent by the French Minister of the Interior to check the authenticity of the reports. Biot's detailed account of the event left the French scientific establishment with no alternative but to accept that stones did sometimes fall to Earth.

The fall at L'Aigle was a landmark in understanding meteorites. Once again, it became fashionable to include meteorites in collections of rocks and minerals. From then on, collections of meteorites began to accumulate in private hands and in the world's major museums. Modern science benefits greatly from this legacy, as preserved fragments of around 1000 witnessed meteorite falls from all over the world are available today for research.

Despite the importance of the L'Aigle meteorite, the most significant early study of meteorites predated the work of Biot. A couple of years before the fall at L'Aigle, Sir Joseph Banks, then president of the Royal Society in England, acquired samples of stony meteorites including pieces of the Siena and Wold

Cottage falls, persuading a young English chemist, Edward Charles Howard, to analyse them.

Supplied with additional stones from other sources, Howard realised that to perform the analyses he had to separate some of the components of meteorites. To achieve this he worked closely with a French nobleman and mineralogist, Jacques-Louis the Comte de Bournon, who had earlier fled to England to escape the terror of the French Revolution.

Most stony meteorites contain abundant grains of metallic iron, and de Bournon separated these for analysis. In the metal, Howard found a considerable amount of nickel, an element that had also been found in an iron meteorite from Argentina by a French chemist, Joséf-Louis Proust. Howard published the work in 1802, and for the first time a chemical link was established between stony and iron meteorites that had fallen at different times in widely separated places on Earth. It was not long before a common, extra-terrestrial origin for meteorites became accepted internationally.

The questions of where meteorites had come from in space and how they had become launched towards Earth, were not answered properly for more than another century and a half. Howard's work was quickly confirmed by the French chemists Louis-Nicholas Vauquelin and Antoine de Fourcroy, and by the German chemist, Martin Heinrich Klaproth. Various scientists speculated that meteorites had come from the Moon, and this view remained entrenched until well into the twentieth century.

METEOROLITES.

In 1801, Ceres, the largest and first to be discovered minor planet (or asteroid), joined the Moon as a possible source of meteorites. Soon the incorrect idea that meteorites (and asteroids) were fragments of a single fragmented planet, represented by debris in the asteroid belt between Mars and Jupiter, took hold and persisted for more than a century.

In spite of these great advances, some die-hards resolutely stuck to the belief that meteorites originated on Earth. Many sceptics of the 'cosmic origin' voiced their opinions well into the nineteenth century. The most notable, Thomas Jefferson, while occupying the office of President of the United States, is said to have greeted reports by scientists at Yale University of the fall of a stony meteorite at Weston, Connecticut in 1807 with the words: 'It is easier to believe that two Yankee Professors would lie than that stones should fall from heaven'.

Whether Jefferson actually uttered these words, or whether they have been embellished in the repeating is disputed, but the sentiment was widely held among the gentry of the day in both America and Europe. In the same year, as the scientific view began to prevail, Afanasii Stoikovich, Professor of Physics at the University of Kharkov in Russia published his book *Of Aerial Stones and their Origin*. Stoikovich reviewed the suggested origins of meteorites, and considered that the hypothesis that they were fragments of asteroids was 'not in contradiction to the laws of physics'.

On the title page of his book Stoikovich included a Latin inscription concerning the Ensisheim meteorite, that had fallen more than three hundred years before. It translates as 'All know about it, each a little, but no-one enough', and these challenging words provided an inspiration to many scientists for the rest of the nineteenth and twentieth centuries.

Today, more than 200 years after Ernst Chladni proposed it, we know that meteorites are fragments of extra-terrestrial material surviving their fall to Earth from space. The accumulated chemical and astronomical evidence confirms that meteorites originate within the Solar System. Most are pieces of rock and/or metal broken from many asteroids in elliptical orbits around the Sun. Having remained essentially unaltered since their formation, modern dating shows that most meteorites are 4555 million years old.

As samples from inactive minor planets, meteorites are our only direct source of information on a wide variety of events linked with the birth and earliest history of the Solar System. The study of mete-

orites has resulted in huge technical and philosophical advances in physics, chemistry, mineralogy and astronomy, providing a crucial link between astrophysics, planetary science and geology.

If the recognition of the scientific importance of meteorites was slow, then overlooking the potential effects of asteroidal impact with the Earth could have dire consequences for humanity. Ancient rocks on Earth bear abundant but cryptic evidence that large projectiles have sometimes struck the planet, excavating huge craters. Occasionally, these collisions may have caused worldwide catastrophes. Remarkably, not until the 1930s and 1940s did many geologists accept that some of the more obvious craters on Earth were formed by meteorite impact.

Realisation that impacts are of global significance came only after the start of the Space Age in the 1960s. And it was not until the 1980s that chemical evidence discovered in sediments of the same age around the world suggested that a huge chunk of Solar System debris had slammed into the Earth around 65 million years ago at the close of the geological period called the Cretaceous. The coincidence of this chance event with the apparent extinction of 70 per cent of the world's animal and plant

The Wold Cottage stony meteorite which fell in Yorkshire in 1795 (shown with other meteorites from Britain).

Monument marking the site of fall of the Wold Cottage meteorite.

A fireball was observed over London's Fulham Road in the nineteenth century.

The Ensisheim meteorite, and the mystery surrounding it and others, inspired Afanasii Stoikovich in his study. Stoikovich proposed that meteorites originated from asteroids.

Meteor Crater in Arizona USA measures 1.2 km wide and 180 m deep. It is now known that it was formed 50 000 years ago by the explosive impact of an iron meteorite weighing more than 50 000 tons. But despite abundant meteoritic fragments littering the rim and surrounding plain, it was not until the 1930s and 1940s that the true origin of the structure was accepted by the scientific community, and it was not until the 1960s that the mechanics of crater formation were understood.

genera sparked serious research into the possibility that catastrophic collisions could have punctuated our geological and biological history.

Our natural satellite, the Moon, may owe its existence to a collision between the infant Earth and a body about a third its size. The impact tilted the Earth on its axis, with the debris from the collision collecting to form the Moon. Far-reaching consequences of this random early event, vital to life on Earth today, include the length of the day, the seasons, and tides.

Those who thought that the Solar System is no longer a dynamic place would have had their doubts removed by the spectacular collisions of the 22 fragments of the Comet Shoemaker–Levy 9 with Jupiter in July 1994. A subject once considered 'esoteric' by some geologists, today meteoritics is at the forefront of scientific thinking. Importantly, the search for asteroids and comets that may pose a threat to civilisation is a matter that concerns us all.

29 July 1773 At half after eight in the evening, the air serene and calm, and the

moon very brilliant, approaching to the North West, a meteor appeared at Crespi,

in France, in the form of a globe, to which was affixed a tail, placed vertically:

the light reflected by it was so considerable as to obscure that of the moon for

some short space; after which the meteor began to decrease in splendour, and

tended towards the earth; and about seven minutes after a noise was heard equal

to the round of the largest cannon, and caused such a commotion as to shake the

glasses and other movable bodies in all the houses. The same meteor was seen

about the same time in Paris, but they heard nothing of the explosion.

THE GENTLEMAN'S MAGAZINE, SEPTEMBER 1773

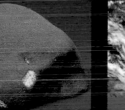
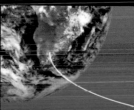

ROCKS FROM SPACE

Although there is no record of a meteorite recovered anywhere near the town of Crespi (now Crepy) in 1773, the accuracy of the description leaves no doubt that what the people of the region witnessed was the fall of a meteorite. When it comes to meteorite falls, observing and not recovering, or not observing at all, is normal rather than exceptional. World-wide, on average, only about five or six meteorites are both seen to fall and quickly recovered each year, although the actual number of falls each weighing more than 10 g may be as high as 42 000 per year over the entire surface of the Earth.

The large difference between these figures is easily explained. Nearly three-quarters of the Earth's surface is covered by water, and less than a third of the remaining land surface (that is approximately one tenth of the Earth's total surface) is densely populated. Consequently, most meteorites are not recovered, with the greatest number falling into the oceans.

When Kathleen Clifton and Theresa Davies set off for Binningup Beach on the glorious coast of Western Australia in 1984, a cosmic event was certainly not on their minds. It was the 30th of September and a typically cloudless Australian spring day. Just after 10 am, as they settled down to sunbathe, they were startled by a whistling noise and a loud thud in the sand nearby. Frightened, Kathleen called her husband Brian. Thinking that his wife had been shot at, Brian ran into the sand hills to find the 'sniper'. Returning a few moments later, he found a black stone, the size of large potato, in a 15-cm pit it had dug in the sand only 4 m from where his wife and her friend had lain. Picking it up he found that it was quite cool. It did not take them long to realise that they had come close to being hit by a meteorite.

Moments before and unseen by the people at Binningup, a brilliant fireball, lasting for only a few seconds, streaked across the sky above the towns of Toodyay and Pinjarra, bursting like a firework into four or five luminous fragments before disappearing. At the same time, two loud thunder-like bangs were heard in the nearby city of Perth. In the coastal towns of Rockingham and Mandurah, buildings shook and windows vibrated for several seconds. During the weeks following the fall, searches of nearby bushland failed to locate any additional fragments.

Records of recovered meteorite falls show that close encounters like that at Binningup are not unusual. Most quickly recovered meteorites fall within tens of metres of a human being, or else damage buildings and cars. There are only two well substantiated records of people being hit by meteorites, and only two recorded fatalities, that of a dog and a horse. Many historical accounts of meteorites reported to have killed people and animals cannot be substantiated.

Meteorites fall at any time on any part of the Earth. Since the historic fall at Ensisheim to the present day, detailed accounts of many eyewitnesses have been recorded. From these observations, as well as more accurate astronomical measurements, and space technology, there is now a very good understanding of the phenomena associated with meteorite falls. Moreover, estimates of the amount of meteoritic material falling to Earth with time, and the size range of this debris, are being refined constantly.

Currently, during each year's orbit of the Sun, the Earth encounters tens of thousands of tons of natural debris from space, of which only some fall to

HM Walker, from Bangor in northern Wales, photographed the fireball which preceded the fall of the Bovedy meteorite in Northern Ireland at 9:22 pm on 25 April 1969. Two masses, the largest 4.95 kg, were recovered some 60 km apart from near Sprucefield and Bovedy.

Kathleen Clifton and Theresa Davies were the lucky sunbakers of Binningup.

n 1984 the Binningup meteorite flopped onto the sand beside two sunbathers on a beach in Western Australia.

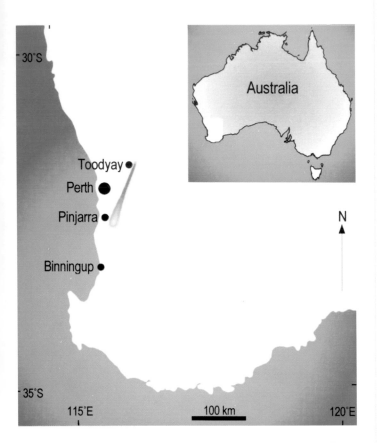

Earth as recoverable meteorites. Many are small fragments that appear as meteors when they burn up high in the Earth's atmosphere. By far the most abundant are microscopic particles of dust, interplanetary dust particles and cosmic spherules, called micrometeorites, that drift unseen to the surface. While in space, all of these small natural objects are called meteoroids or micrometeoroids.

'Dust' generally covers particles less than 1 mm across, or about the size of a grain of salt. Because of their large surface areas compared to their tiny masses, some dust particles can quickly radiate back the heat generated from friction with the atmosphere, and so fall unmelted, and their descent is not accompanied by light. However, dust also includes droplets stripped from the surfaces of larger fragments and micrometeoroids flash-melted during their atmospheric passage.

Another category of dust, interstellar dust, deserves a special mention. Regions between the stars contain gas and dust. This 'interstellar medium' is the seeding ground for new stars. Although it has long been suspected that interstellar dust enters the Solar System, the size and amount of material making its way to the inner Solar System, and possibly to Earth, has evaded detection until recently. Dust-detecting spacecraft and astronomical measurements show that interstellar dust particles are usually less than a thousandth of a millimetre in diameter. As we shall see later, some meteorites contain even smaller grains from beyond the Solar System, but interstellar meteorites or micrometeorites, if they exist, have yet to be recognised.

Earth's gravity acts like a giant lens, focusing material in its near vicinity to capture matter from a region wider than its diameter. Although meteorites fall randomly, this gravitational effect results in a slightly greater number of falls near the equator compared with the poles.

The time meteorites fall gives a clue to the way they move relative to the Earth. Records show that there are many more witnessed falls during the day

Left: The flight path of the Binningup meteorite can be reconstructed from sightings of its fireball over Toodyay and Pinjarra.

Only two people have been hit by meteorites in modern history, one in Uganda on 14 August 1992. Although a large shower totalling 108 kg fell over the densely populated area around Mbale, and several buildings were hit, nobody was hurt: only one boy was hit on the head by a small stone that passed through the canopy of trees.

At 6:30 pm on 18 August 1974, a 2.7-kg stony meteorite fell through the roof of a school in the village of Naragh in Iran. The hole in the roof (top) measured about 90 cm in diameter. Fortunately the science laboratory (below) was unoccupied when the meteorite fell.

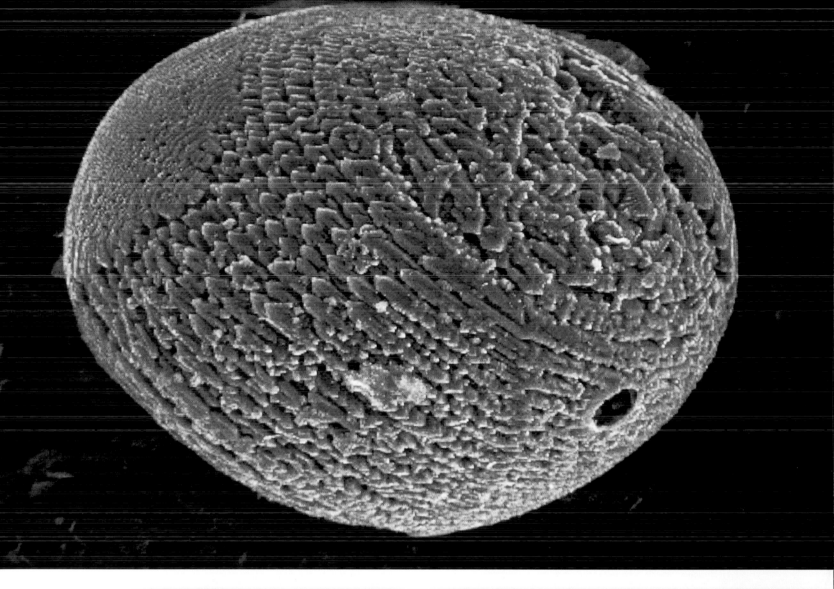

Meteoritic dust, such as this melted micrometeorite picked up from Antarctic ice, accounts for the greatest amount of extra-terrestrial material falling to Earth each year. (About 0.5mm across).

Interplanetary dust particles, measuring around one hundredth of a millimetre across, are aggregates of many tiny mineral grains. Their sources are unknown, but they may originate from comets.

20KV X7200 0138 1.0U OCT82

than at night. The obvious conclusion is that people are more active during the day than at night. A closer look at meteorite falls during daylight hours, however, shows that there are about twice as many in the afternoon and evening, compared with the morning. The motion of the Earth, rather than human bias, provides the explanation.

Because the Earth spins on its axis, for local times between noon and midnight, the half of the Earth facing away from the direction in which it is moving is struck by debris moving in the same direction as the Earth. In this case, it is more likely that gravitational attraction will capture the debris, the low approach speeds of small meteoroids relative to the Earth favouring their survival.

Between midnight and midday, the half of the Earth facing forward meets debris head on. Colliding at speeds equal to the sum of the Earth's and the meteoroids', few objects survive. The occasional meteorites that fall at night or early morning are those that are overtaken by the Earth, or where gravitational focusing has caused a meteorite to fall on the side of the Earth away from the meteoroid's direction of motion. The majority of meteoroids then must circle the Sun in the same direction as the Earth, with most of those surviving to fall as meteorites hitting the Earth from behind while overtaking it.

Even if a meteoroid has no motion relative to the Earth, the planet's gravitational attraction causes it to enter the atmosphere at a minimum speed of 11.2 km/s (or 40 320 km/h), which is about 40 times faster than the speed of sound. This is the Earth's 'escape velocity': that is, in reverse, the speed a rocket has to reach to overcome the gravity pulling it back on to the surface. At such high speeds, frictional heating by rubbing against atmospheric molecules causes the surface of the body to melt and vaporise, and a large volume of the air surrounding the object to become electrically charged, or ionised. If objects are large enough to penetrate deep into the atmosphere, the resulting phenomenon of incandescent gas and dust, called a fireball, gives rise to brief, spectacular visual displays and sometimes terrifying noises. An older term, 'bolide' (from the Greek *bolis* meaning 'missile'), is sometimes used to describe large fireballs that explode with loud bangs.

Light and sounds accompanying meteorite falls are often quite startling, but what is it that is actually happening? During high-speed flight through the atmosphere, melted material is stripped away rapidly from the surface of a meteoroid, to produce a trail of incandescent gas and solidified droplets. This stripping process, called 'ablation',

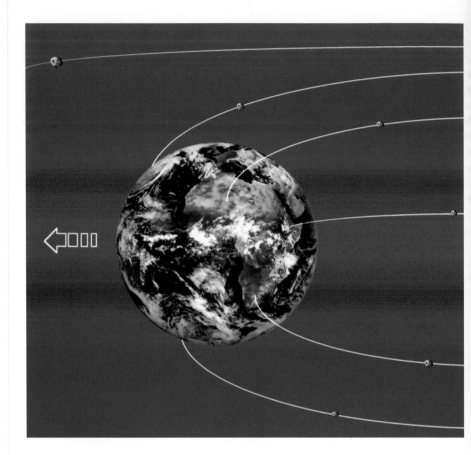

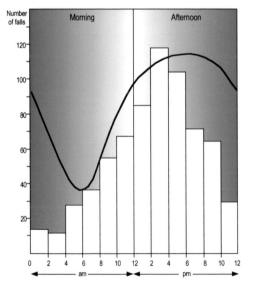

The numbers of falls, from a total of over 600, recorded at given hours of the day or night show that there are twice as many in the afternoon as in the morning. The line curve was calculated from trails of possible falls, and is independent of human observation, showing that there is a real variation in the diurnal distribution of falls caused by the Earth's motion.

The Earth's gravity attracts meteoroids from a region of space wider than its diameter, and 'focuses' especially small, slower objects towards the equator. Sometimes the gravitational pull can even cause a meteoroid to pass around the Earth and fall on the opposite side.

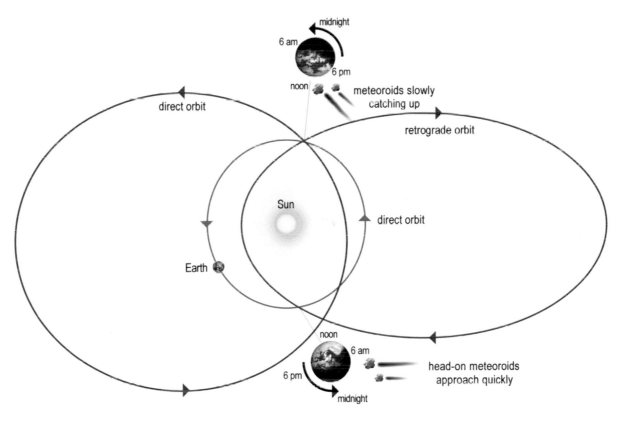

The diurnal variation of meteorite falls is explained by the rotation and orbit of the Earth relative to the orbits of meteoroids. Meteoroids with orbits in the same (direct) sense as Earth have slower relative speeds and are more likely to fall in the afternoon and survive as meteorites. Faster meteoroids orbiting in the opposite (retrograde) sense occur mainly at night and most are destroyed. The bias in the times of observed meteorite falls shows that most come from material orbiting the Sun in the same direction as Earth.

When debris from space enters the atmosphere, the tiniest particles are destroyed in the upper atmosphere, and appear as meteors. Larger objects generate fireballs and survive to lower altitudes above the surface, but may also be destroyed. Most fireballs extinguish at 10-30 km. Objects with low relative speeds are slowed by the atmosphere and fall under gravity, sometimes breaking up to produce meteorite showers. Large projectiles are not significantly slowed by the atmosphere and can cause explosion craters.

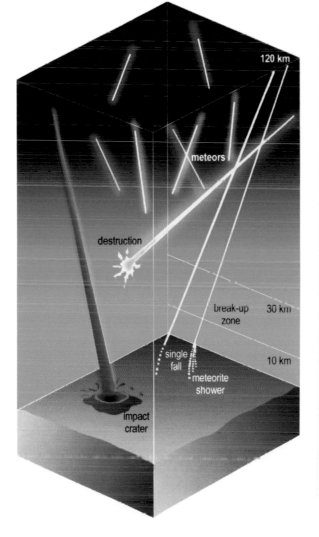

removes melted material so efficiently from the meteoroid that heat is generally not conducted into its interior, which remains at the freezing temperatures of outer space. Objects large enough to survive are usually slowed by friction in the dense lower atmosphere, to a height (commonly 10-30 km above the Earth's surface) where frictional melting ends, and the fireball extinguishes. Heating, melting and ablation of meteoroids in the atmosphere usually lasts from a few seconds to a few tens of seconds. In the last second of luminous flight, the molten surface solidifies to form a thin 'fusion' crust, and the body, with no remaining high 'cosmic' speed, free-falls to the surface under the normal effects of gravity to become a meteorite.

During a meteorite's free-fall to Earth, its fusion crust cools rapidly. Even when they are picked up immediately, most freshly fallen meteorites are rarely more than lukewarm to the touch. In rare cases, because of their icy cold interiors, a feathery white 'hoar frost' or condensation forms on the surfaces of freshly fallen meteorites. Historical records of 'hot' meteorites cannot be substantiated, and meteorites certainly do not set fire to the objects they hit. Some freshly fallen iron meteorites, however, have been reported to be quite warm. This could be explained by the ease with which metallic iron conducts heat compared with rocky materials. Nevertheless, the effects of frictional heating can be seen in less than the outermost centimetre of most meteorites.

Although a meteoroid may be the size of a basketball, the fireball that it generates in the lower atmosphere can be several hundred metres across. Many eyewitnesses have described fireballs that appeared to them as the size of the Sun or Moon. Fireballs sometimes leave large smoke trails made of tiny particles in their wake and, in favourable conditions at night, a luminous cloud of weakly ionised atmosphere can persist for a long time after the passage of a fireball.

Meteorite falls are accompanied by sounds that vary considerably, although these are not always heard by observers. In the same way as a supersonic aircraft, or high-velocity bullet, meteoritic bodies compress the air in front of them creating shock-waves in the atmosphere. In the denser lower atmosphere, as bodies are slowed through the sound barrier, shock-waves generate 'sonic booms'. At least two or three loud 'bangs' are often heard, invariably followed by sounds like the rumbling of a distant thunderstorm.

Even more alarming are so-called 'electrophonic sounds' like hissing, sizzling and buzzing noises that are heard while the fireball is visible. Likened to bacon frying in a pan, electrophonic sounds are of low intensity and not everyone can hear them. Because their frequencies are in the range of the activity of the brain, some people 'sense' these sounds rather than hear them. Animals are particularly sensitive to electrophonic sounds, and in documented cases have shown alarm shortly before a meteorite fall. One explanation to account for these strange sounds that travel apparently at the speed of light is that they are low frequency radio transmissions from electromagnetic radiation generated by the fireball.

Some meteoroids can lose up to 96 per cent of their initial mass as a result of ablation, while others lose very little. Clearly, the size of the body before it enters the atmosphere is important for survival, but the nature of the material also plays a role. Meteoroids composed of weak, crumbly materials that are easily disintegrated stand less chance of survival than strong, compact bodies. The initial speed and shape of a meteoroid combined with the angle of trajectory in the atmosphere are also important factors determining survival. Most meteoroids enter the Earth's atmosphere at speeds of 15-70 km/s, although the average speed of those that survive to become meteorites is around 17 km/s. At the upper end of the range, above about 30-40 km/s, objects are destroyed on impact with the atmosphere. During such high-speed encounters, air can act as if it is a solid.

Above: The Binningup meteorite is covered with a thin (less than 1 mm) fusion crust which formed during its passage through the atmosphere. The white interior remained unaltered by the heat that melted the crust.

Fusion crusts on meteorites (Camel Donga, left; Binningup, right) can vary considerably in texture and appearance. The first crust to form is generally smooth, although a secondary crust can form over irregular broken surfaces.

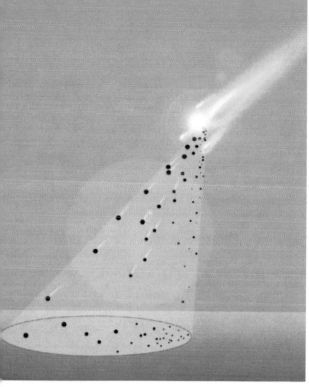

Above: When meteorites break up into showers, the larger fragments are carried further by their greater momentum, producing a characteristic elliptical distribution.

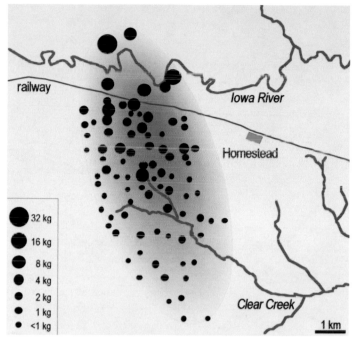

Right: Over 100 stones weighing together about 227 kg were recovered from a classic elliptical area of around 46 km² from a stony meteorite fall in 1875 near Homestead, Iowa USA.

Observations of actual meteorite falls suggest that the strength of the material is crucial. As a result of the pressure of air and the resultant stresses, meteoritic bodies commonly break up in the atmosphere. In fact, falls of single meteorites are rare. Large, brittle stony meteoroids often shatter during atmospheric flight, releasing showers of meteorites comprising tens, hundreds, or even tens of thousands of individual pieces.

The meteorite shower with the largest number of pieces yet recorded fell at Pultusk in Poland on 30 January 1868. Estimates put the number of fragments in this shower of stony meteorites at around 100 000, many of which were pea-sized, and altogether totalling approximately 2 tons. The shower with the greatest total weight, estimated at around 70 tons but from which some 23 tons was collected, fell on 12 February 1947 at Sikhote-Alin in Siberia. The Sikhote-Alin meteorite is the largest observed fall of meteoritic iron on record.

Depending on how high in the atmosphere and at what velocity meteoroids fragment into showers, each fragment can develop its own fusion crust. The first crust to form is called 'primary' crust, but a thinner, 'secondary', crust can coat fractured surfaces. Clattering and machine-gun like sounds sometimes heard during the fall of meteorite showers are due to the break-up of the meteoroid into smaller pieces, each with its own sound wave.

When meteorites break in the atmosphere to give rise to showers, a cone of fragments spreads out, with each fragment having an independent flight to Earth. The intersection of the cone of fragments with the surface of the Earth produces an elliptical distribution of meteorites on the ground. So-called 'strewn fields' of meteorite showers vary in length and width, sometimes being over 50 km long. Because of their greater momentum, large fragments usually travel further than the small fragments in a shower. Within a strewn field there is often a gradation in size from small to large meteorites along the direction of fall.

The largest impact pit from the Sikhote-Alin shower in Siberia in 1947.

OBSERVING A
METEORITE FALL

Two of the most popular misconceptions held by witnesses of meteorite falls or large fireballs are firstly, that the fireball continues to the ground so that it's intersection with the observer's horizon is where the meteorite fell; and secondly, that the loud noises are the meteorite hitting the ground. Observing transient events in the sky, like fireballs, can be very deceptive because there is nothing against which to gauge our observations.

When we throw or catch a ball, for example, our eyes and brains co-ordinate to determine the path of the ball very accurately. We are able to do this because humans see in three dimensions with binocular vision, and our brains automatically calculate how far away, how fast and in what direction the ball is moving in relation to the ground, a fence, a house, a tree; indeed all of the stationary structures around us. When we observe an object moving through the atmosphere, our eyes have no such fixed reference points of known size. Consequently, there is a tendency to overestimate its size. Save for one observation, it is extremely difficult for a single observer to judge accurately the direction in which a fireball is travelling. The exception is when an observer happens to be looking directly along the flight path. In this case the fireball does not move much at all. Instead it would appear stationary or moving very slowly, grow larger and more intense, perhaps wobble a bit, then disappear.

Most observations of meteorite falls are not made by scientists. Witnesses near the point of impact of a meteorite are left in no doubt that something unusual has happened. Attention is often drawn to the event by the 'rushing' or 'whistling' noise accompanying a high velocity projectile as it passes close by, or the 'thud' of the meteorite hitting the ground.

If you are fortunate enough to witness a fireball you can make a significant contribution to science by recording a few simple observations.

NOTE
* your exact position when your observations were made

* the exact time the fireball appeared and how long it lasted

* the directions, and angles from the horizontal of your first and last views of the fireball relative to any prominent landmarks. (Sketching your observations can help. At night, note the path of the fireball with reference to the stars.)

* what angle the path of the fireball made with the horizon

* how bright, and what size and shape the fireball was

* did it have any colour

* did you hear any sounds

* did the fireball leave any visible trail

* whether the fireball exploded

* how long the luminous trail remained in the sky.

Reports of this kind of information from many observers with different views of a fireball would enable scientists to reconstruct an approximate path in the atmosphere and so predict where any surviving meteorites might be found.

When reporting the discovery of a meteorite, details of the exact location of the find can be important. If the meteorite made a depression, or became plugged in a hole on impact, note the shape of the scar, or inclination of the hole relative to points of the compass. Even if no fireball was seen, this kind of information could help scientists to determine the flight direction of the object. Above all, avoid contamination of the meteorite by excessive handling. Place it in a clean, sealed polythene bag. De-airing the bag and placing it in the ice-box of a refrigerator may help preserve some of the meteorite's short-lived, temperature-sensitive properties.

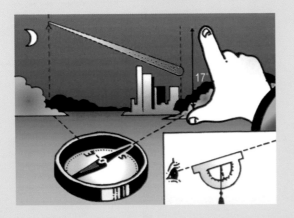

The accurate observation of a fireball can help locate any surviving meteorites. Important measurements are the directions and heights of the start and the end of the fireball's trial. Note your exact position when making measurements. You can measure angles using a protractor, or a natural gauge made between the forefinger and thumb at full arm extension, which measures 17° in everyone.

The largest known meteorite in the world, the Hoba iron, remains in place where it was found in Namibia, Africa. Dr Candace Kohl, University of California, provides a scale for comparison.

Shapes and sizes

Meteorites vary in weight from fractions of a gram to tens of tons. Sculpting of their surfaces by ablation in the atmosphere and subsequent weathering on Earth produces a great variety of shapes. When meteoroids point in the same direction during atmospheric flight, then their leading surfaces are ablated uniformly resulting in cone, or so-called 'orientated', shapes. Meteorites that tumble, or are fractured in the atmosphere, generally have more irregular shapes.

Fusion crusts on freshly fallen meteorites vary enormously. Often smooth, crusts can also be decorated with spattered droplets, or strings and rivulets of molten material. Rounded pits and depressions resembling thumb prints in a ball of clay are found on the surfaces of many meteorites. Called regmaglypts (from the Greek *regma* meaning 'cleft' and *glypto* meaning 'to cut'), these atmospheric sculptures formed during ablative flight where swirling air over the surface of a meteoroid caused uneven melting. On orientated meteorites, regmaglypts radiate from the apex of the cone towards the rear surface, indicating the flow of compressed air.

Considering their high initial speeds, the great majority of meteorites inflict remarkably little damage. The largest single meteorite known was found in 1920 near Grootfontein in Namibia, in south-west Africa. This iron meteorite, named Hoba, weighs approximately 60 tons and remains to the present day as a national monument at its find-site. Meteorites up to ten or more tons, such as Hoba, may bury themselves in deep impact pits. In the great majority of cases, neither the meteorites nor the ground on which they fall suffers much damage.

Excavated from the 6-m hole it punched in soft sediment, at 1.77 tons the largest known single stony meteorite fell as part of a shower at Jilin, China on 8 March 1976. Depending on the nature of the ground, smaller meteorites also plug themselves in shallow holes that may be slightly inclined in the direction of motion of the meteorite. The largest single fragment of the Sikhote-Alin shower, weighing 1.75 tons, was recovered from the bottom of a small impact pit 4 m deep where material had been thrown up in all directions. Broken into smaller pieces on impact, the largest mass of this shower dug a crater 26 m in diameter.

In a remarkable photo of a fireball taken over Australia at 1:45 am, 1 May 1995 from the window of a truck (the light cast on the frame of the window bottom right), the fireball appears to corkscrew through the air. Unfortunately no meteorite was recovered.

The Bunjil stony meteorite (right) was uncovered in Australia in 1971 during ploughing. Its regular shape resulted from uniform ablation in the atmosphere. The deep regmaglypts of the Haig iron meteorite (left) formed by uneven atmospheric melting of the surface.

Irrespective of their initial speeds, similarly sized small masses generally result in the same effects when they hit the surface. The reason for this is the equalising nature of air drag in the atmosphere. The higher the speed of an object the greater the air resistance. Above the speed of sound (1118 km/h) air resistance becomes four times greater each time the speed doubles. Consequently, meteoroids with higher initial speeds are slowed more quickly in the atmosphere and have a greater chance of destruction than those with low speeds. After the point where all of its cosmic speed has been removed, a surviving object will reach a terminal speed determined by gravity and air resistance. Typical terminal speeds for free-falling objects of 1-100 kg are 125-250 m/s, and this would be achieved within a few seconds after luminous flight ends.

Larger meteoroids weighing thousands of tons, more massive than the column of air they displace, are not slowed appreciably by atmospheric friction. These projectiles retain most of their initial speed. In their case, a fireball would persist to the ground and the release of energy on impact would result in an explosion exceeding that of a nuclear bomb. Meteorite impact craters, evidence of the devastating effects of such collisions, are found at many places on Earth.

Meteorights and meteowrongs

Meteorites come in all shapes and sizes and are made of quite diverse materials. In most parts of the world the chances of finding a meteorite are remote. Meteorites are usually recognised because they look very different from other rocks in the same area, and so appear 'exotic' to the finder. However, there are many natural and man-made materials commonly mistaken for meteorites. Perhaps the most abundant and widespread objects masquerading as meteorites are rounded nodules of ironstone.

Ironstone is a general term used to describe a variety of naturally occurring rocks and concretions formed over millions of years on Earth. As their name suggests, ironstone nodules are rich in iron, in the form of iron oxides giving them a rusty appearance. Concretions of ironstone form by the deposition of iron from groundwaters seeping through rocks. Often harder than their surroundings, ironstone concretions frequently survive after the rocks in which they formed have completely eroded away, leaving them lying on apparently unrelated surfaces.

Other objects commonly mistaken for meteorites are near-spherical nodules of the mineral pyrite (iron disulfide) that also form concretions in some rocks. Despite its metallic appearance, pyrite is

The largest mass of the Jilin stony meteorite, which fell in China on 8 March 1976, was recovered from its impact hole (left) after extensive excavation (right).

Ironstone and pyrite nodules (top and middle right), which form naturally in Earth rocks, and industrial waste (bottom right), are often mistaken for meteorites (left).

richer in sulfur than in iron and gives off a smell of burning sulfur when hit with a hammer. Weathered nodules have brown crusts of iron oxides, but on freshly broken surfaces they show long, radiating brassy yellow crystals often ending in crystal faces.

Other common pseudo-meteorites include a variety of rock types, and the products and wastes ('slags') from industrial processes. Industrial slags, made of a variety of materials, such as cast-iron, glass and sil-

icates, are the most convincing of pseudo-meteorites. Rounded balls of manufactured cast-iron used in grinding mills are also commonly mistaken for iron meteorites. Common features of furnace wastes are bubbles or cavities formed by the expulsion of gases during melting and solidification. With only a few exceptions, meteorites never contain gas bubbles.

What are the common features of genuine meteorites that set them aside from rocks on Earth?

The wing-shaped Mount Dooling meteorite found in Australia in 1979 weighs 701 kg.

The Wolfe Creek Crater in the Kimberley of north-Western Australia was formed around 300 000 years ago by the impact of an iron meteorite.

Freshly fallen stony meteorites are coated with matt, or sometimes shiny, black fusion crusts no more than a few sheets of paper thick. Their rocky interiors may be exposed where the fragile crusts have peeled away. Because of the diversity of meteorites, fusion crusts vary considerably in appearance. Spotting deeply weathered, or rare types of meteorites requires an experienced eye. A few meteorites that are exceptionally poor in iron bear crusts made of clear glass, the rocks appearing creamy white.

Most meteorites contain some metallic iron, readily attracting the needle of a compass. But their magnetic properties vary with the amount of metallic iron they contain. In fresh examples of common stony meteorites, metal shows up on cut surfaces as abundant silvery specks. Meteoritic metal always contains nickel, but contains very little carbon. This

is very different from industrial cast iron, that lacks nickel and is made mainly of iron and carbon. Commercial steels can contain nickel, but may also contain substantial amounts of manganese, chromium and tungsten that meteoritic metal does not.

Because of their generally high contents of iron, meteorites are heavy objects. Iron meteorites weigh about three times as much as common Earth rocks of comparable size, and stony meteorites about one-and-a-half times as much. However, not all meteorites share these characteristics and there are some terrestrial rocks that rarely contain metal and have the outward appearance of meteorites. Ultimately, meteorite identification requires expert examination and we encourage people who suspect they may have found a meteorite to contact geologists at their local museum or university.

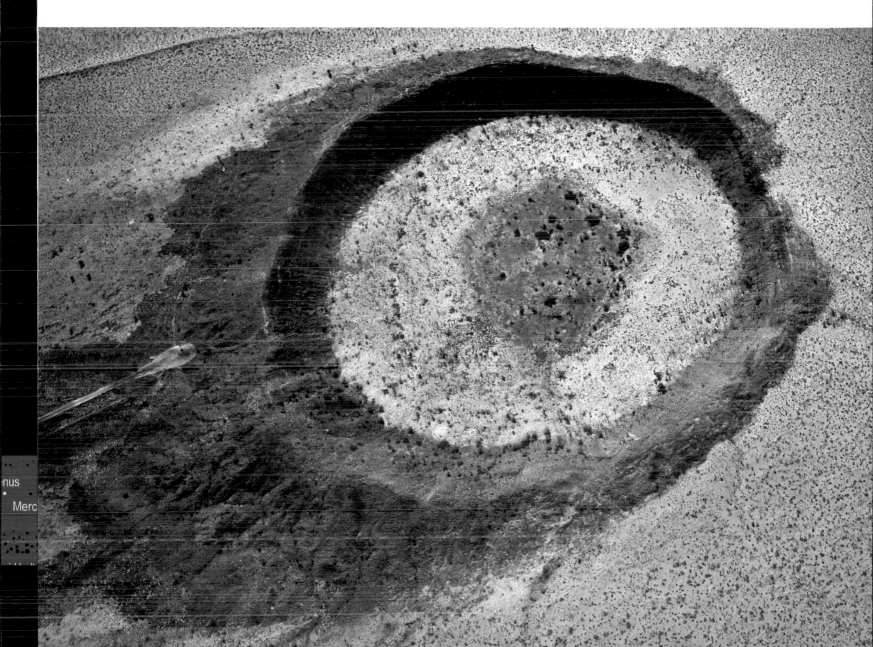

CELESTIAL VOYAGERS

Where do meteorites come from? The asteroids are prime suspects, but what evidence is there to support this, and do *all* meteorites and micrometeorites come from asteroids? The answer to the latter question lies in chemical and other evidence presented by some meteorites, interplanetary dust particles and micrometeorites, and is dealt with in detail later. That the majority of meteorites come from asteroids is supported by measurements.

To identify the likely sources of the meteoroids giving rise to meteorites, their paths in space relative to the Sun (orbits) have to be calculated. To do this, the exact time of the appearance of a fireball has to be recorded, a measurement of its initial speed (that is, of the meteoroid causing it) has to be made, and its path through the atmosphere determined.

Meteorite falls are unpredictable, and momentary glimpses of fireballs are rarely accurate enough to use for calculations. A reasonably accurate time of entry could be gained from observations by many witnesses, as could a rough flight direction. The initial speed of a meteoroid is the most difficult measurement to obtain. To measure this the motion of the Earth, and other bodies in the Solar System, around the Sun have to be taken into account.

Following in the footsteps of the great Danish astronomer Tycho Brahe (1546-1601) who measured planetary motions, in 1609 the German physicist Johannes Kepler (1571-1630) formulated the cardinal principles of modern Solar System astronomy. Kepler showed that the planets move around the Sun in slightly egg-shaped, or elliptical orbits, the shape of which is determined by the average distance from the Sun, and the focus in the ellipse where the Sun is situated.

Imagine playing totem tennis. Instead of the ball swinging freely in a circle around its pivot, the string catches and becomes shorter. As the string shortens the ball speeds up as it gets closer to the totem. Freeing itself, the string becomes longer again and the ball travels around the totem more slowly. By our analogy, if the pivot of the totem represents the Sun, then for an elliptical orbit the length of the string when the ball is on one side of the totem is longer than when it is on the other side, and there is a constant change in the length of the string as the ball travels around it.

Consequently, the speed of an object in an elliptical orbit varies, being greatest nearest the Sun, and least at its farthest point from the Sun. Essentially, the speed along its orbit decreases slightly as the distance between the object and the Sun increases. Isaac Newton (1642-1727) later provided the theory for Kepler's laws by formulating the Law of Universal Gravitation whereby every particle of matter attracts every other particle with a force varying both with their respective masses, and how far they are apart.

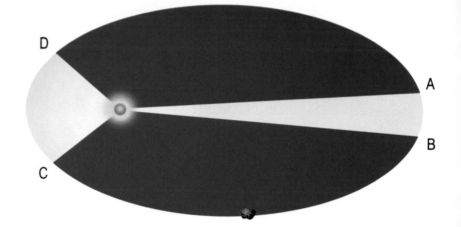

Johannes Kepler showed that for an object in an elliptical orbit around the Sun, the areas of space swept by an imaginary line joining the body to the Sun are equal for the same time interval. Thus at its greatest distance from the Sun, the body moves more slowly (from A to B) than it does when closest to the Sun (from C to D).

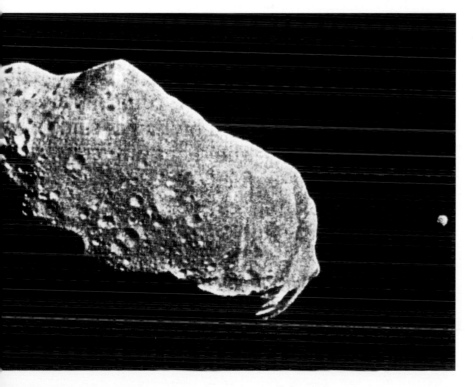

The Earth revolves around the Sun in an almost circular orbit at an average speed of 29 km/s. To obtain the actual speed of a meteoroid relative to the Earth we must subtract from its measured speed the Earth's escape velocity of 11.2 km/s, caused by its gravitational pull. A small correction also has to be made for the speed the Earth spins on its axis, which at the equator is just less than 0.5 km/s. The combination of the meteoroid's speed relative to the Earth, and the Earth's own speed around the Sun, gives us the speed of the impacting body relative to the Sun. Using our totem tennis analogy, at the moment of impact with our atmosphere the 'length of the string' for the meteoroid equals the distance between the Sun and the Earth. From this, the time that the fireball appeared, and from its speed, the dimensions of the meteoroid's orbit can be calculated.

By painstakingly processing reports of eyewitnesses, using mathematical 'trial and error' to fit the initial speeds of bodies entering the atmosphere, the approximate orbits for only a few meteorite falls had been calculated up until the early 1950s. The most significant was the orbit of the huge Sikhote-Alin meteorite. Russian scientists calculated that the parent meteoroid of Sikhote-Alin had a highly elliptical orbit with its most distant point from the Sun lying within the belt of asteroids between Mars and Jupiter. This was strong evidence, albeit circumstantial, that meteorites were fragments of asteroids.

The asteroids

While some asteroids orbit the Sun inside the Earth's orbit, and others have highly elliptical orbits that cross the orbit of the Earth, the largest number of asteroids are in nearly circular orbits around the Sun in a broad zone, or belt, some 5.5 million km wide, between Mars and Jupiter.

The largest known asteroid, Ceres, is about 933 km in diameter. Since they were first recognised, the convention is to number asteroids, so the discovery of 1 Ceres was followed by 2 Pallas (523 km), 3 Juno (244 km) and 4 Vesta (530 km), and so on. Today, there are more than 4000 named asteroids. Estimates put the actual numbers of asteroids in the tens of thousands, most of which are much smaller than 100 km in diameter, together with countless smaller bodies and fragments.

OUR PLACE
IN SPACE

Looking out from Earth, there are thousands of millions of objects around us in space. These are not arranged randomly, but are gathered into various groups and clusters. The most obvious group is the Solar System; our star, the Sun, and the eight planets that revolve around it. Looking beyond the Solar System, on a clear night with unaided eyes up to 2000 or so stars can be seen; with binoculars about 100 000; using a large telescope more than a billion stars can be detected.

Overhead, stretching across the sky, there is a hazy band of light called the Milky Way. The Milky Way is a clue to the existence of larger groups of stars. Our Sun and its system of planets, and about 100 billion other stars, are arranged into a disk-like mass called a galaxy. The disk, that also contains vast clouds of gases and dust called nebulae, has a spiral structure and rotates slowly like an enormous wheel. The Solar System is situated in one spoke of the wheel about two-thirds of the way from the hub. When you look at the Milky Way, you are looking along the plane of the galactic disc to which our planetary system belongs. It takes one year for the Earth to orbit the Sun, but a single rotation of the galaxy takes about 240 million years.

Viewing just the Solar System, the arrangement of the Sun and planets is also like a rotating disk, with the Sun at the hub. All of the planets move in the same direction around the Sun and in roughly the same plane. In addition, most planets spin on their own axes at various rates. Where planets have natural satellites, or moons, they show similar patterns of movement as the whole Solar System.

Looking at the planets themselves, there are two types. The four planets closest to the Sun (Mercury, Venus, Earth and Mars) are relatively small and dense. The outer planets (Jupiter, Saturn, Uranus and Neptune) are very much larger, but less than a quarter as dense as the inner planets. The outermost body, Pluto, is a small ice-rock planetesimal and is not considered a true planet. It has an orbit that is markedly inclined to the planets, occasionally swinging it between the orbits of Neptune and Uranus.

The distribution of size and density in the Solar System reflects chemical differences in the make-up of the inner and outer planets. The smaller, dense inner planets are mostly rocky and metallic, while the density of the huge outer planets is low because they are made up mostly of gases such as hydrogen and helium, and ices of other light compounds such as water, methane and ammonia. The Sun is around a thousand times heavier than all the planets put together, but because the Sun rotates slowly the planets have more than 98 per cent of the rotational energy of the entire Solar System.

The Messier 83 spiral galaxy, situated 20 million light years from our Solar System, has a spiral structure similar to our own Milky Way.

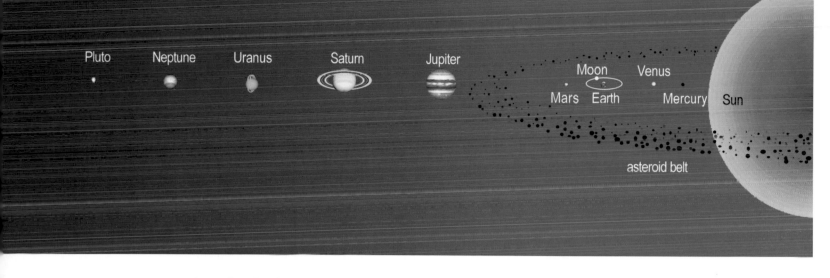

Pluto · Neptune · Uranus · Saturn · Jupiter · Moon Venus · Mars Earth Mercury Sun

asteroid belt

Lumped together this flotsam would make an object only about a twentieth the size of the Moon.

Generations of astronomers have detected three main groups of asteroids. The Near Earth Asteroids include the Apollo, Aten, and Amor asteroids. The Atens are related to their bigger siblings the Apollos, but have orbits inside that of the Earth. The orbits of the Amor asteroids cross that of Mars but only approach the orbit of the Earth, whereas the orbits of the Apollos cross the Earth's orbit.

The second major group are called Main Belt asteroids because they are between Mars and Jupiter in generally stable circular orbits. Main Belt asteroids may include the Hungarias on the Mars side of the Belt, and the Hildas that lie just outside the Belt on the Jupiter side. The third group, called the Trojans, lie ahead and behind the orbit of Jupiter. Another Trojan asteroid, 1990 MB, trails the orbit of Mars.

Within the Main Belt, asteroids are not distributed randomly but occur in narrow rings separated by zones where they are scarce. These generally barren regions are called the Kirkwood Gaps, lying at points of vibration in tune with Jupiter's gravity, where objects take three-sevenths, two-fifths, a quarter, a third, and half the time to orbit the Sun than Jupiter. Asteroids may be in regular orbits for 100 000 years before straying, perhaps nudged by collisions, into the Kirkwood Gaps where they are accelerated and ejected into highly elliptical orbits. If the resulting orbit crosses that of the Earth then the body, or bits of it, may hit the Earth. This makes the Apollo asteroids prime suspects for the sources of many meteorites that strike the Earth today.

Sources of meteorites

Accurate speed measurements for meteoroids in the atmosphere can be made using special cameras triggered by light. Since the late 1950s, sky photography and sophisticated radar techniques have obtained accurate measurements of the speed of bright meteors and fireballs. Cameras with rotating shutters allow the speed of an incoming object to be measured. By accurately controlling the rotation of the shutters, the image of a fast moving object, like a fireball, is chopped into timed segments. Measurements of the height intervals over which a fireball is photographed, combined with the distance travelled in a given time allows the speed at different heights along the trajectory to be calculated.

At 7.30 pm on 17 April 1959, the path of a bright fireball was photographed by a synchronised network of cameras from two stations established to scan the skies over the former Czechoslovakia. Light from the fireball was so intense that the photographic plates were badly over-exposed. Nevertheless, from the salvaged record measurements were made of the speed of the fireball (17.5 km/s), its direction and the position and height of the point in the sky where the fireball extinguished.

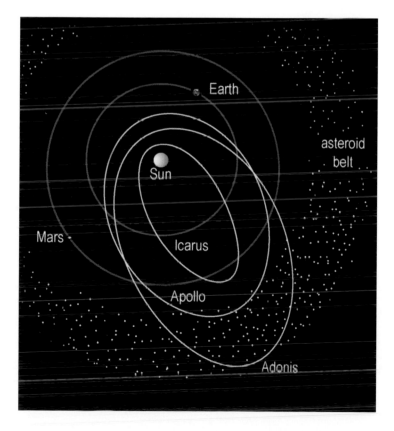

The meteoroid had broken into about 19 pieces in the atmosphere, the photographic record pointing to an area near the Czechoslovakian town of Pribram where meteorites may have landed. Careful searching by volunteers was rewarded with the recovery of four pieces of stony meteorite weighing a total of 5.8 kg. As important as the recovery of the meteorites, for the first time, the photographs allowed astronomers to determine precisely the path that the body followed on its journey through the Solar System to Earth. Calculations revealed that the body had followed an elliptical orbit extending to within the asteroid belt. The Pribram fall confirmed that some meteorites were indeed fragments from asteroids.

Following the success of the All Sky Camera Network in Europe, multi-station networks of cameras were established on the prairie states of the USA in 1965, and in western Canada in 1969.

Disappointingly, in the 40 years since the fall at Pribram, despite the hundreds of fireballs recorded photographically, only two more meteorites have been recovered as a result of All Sky photography.

At 8.14 pm on 3 January 1970, the Prairie Network recorded the fall of a stony meteorite, four fragments of which (totalling 17 kg) were recovered near the town of Lost City in Oklahoma. On 5 February 1977, photographs from the Canadian Meteorite Observation and Recovery Programme led to the successful recovery of stony meteorites weighing 4.58 kg at Innisfree, in Alberta. As in the case of Pribram, the bodies that gave rise to the Lost City and Innisfree meteorites had highly elliptical orbits originating beyond the orbit of Mars within the asteroid belt, but not as far as Jupiter's orbit. Although the orbits crossed that of Earth they did not penetrate as far as the orbit of Venus in the inner Solar System.

The all-sky camera (below) at Streitheim, Germany, forms one part of the Fireball Network, intended to photograph fireballs such as that photographed by the network's camera at Wendelstein, Germany, on 5 December 2000 (left). A rotating shutter chopped the image into 12.5 segments per second, allowing accurate determination of the speed. Its initial mass was estimated at around 270 kg, and a terminal mass of several hundred grams was predicted to have fallen as a meteorite near the village of Mariazell, Austria, but was not recovered.

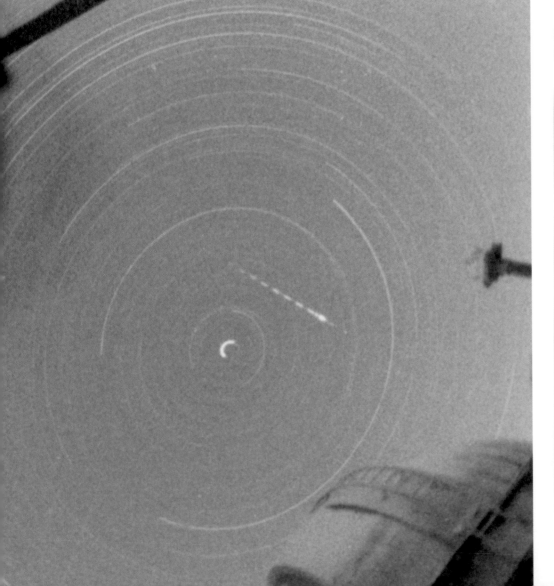

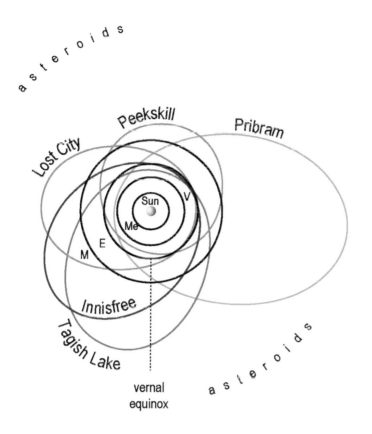

asteroids

Peekskill   Pribram

Lost City

Sun   V

Me

M   E

Innisfree

Tagish Lake

asteroids

vernal
equinox

The calculated orbits
of the bodies that gave
rise to the Pribram
(Czechoslovakia), Lost
City and Peekskill
(USA), Innisfree and
Tagish Lake (Canada)
meteorites all extend to
the asteroid belt
between Mars and
Jupiter, providing evi-
dence that most mete-
orites are fragments
broken from asteroids.

Photographs of the Lost
City fireball on 3
January 1970 enabled
scientists to pin-point
the area in Cherokee
County, Oklahoma,
where the fragments
(three interlocking
shown) landed, aiding
their recovery.

Another two recent meteorite falls were recorded on opportunistic photographs and the information used to estimate their orbits. These were at
Peekskill in the USA in 1992, and Tagish Lake in
Canada in 2000. Their orbits are also elliptical and
indicate the asteroid belt as the source.

Today, of the three large camera networks, only
the European network — now covering the Czech
and Slovak Republics, Germany, Belgium,
Switzerland and Austria — continues to operate. A
wealth of photographic measurements on hundreds
of bright meteors and fireballs, and the three recovered meteorites, show that few were travelling at
speeds greater than the escape velocity of the Solar
System (42.1 km/s at the Earth's distance from the
Sun) and the exceptions could be accounted for by
errors in measurements. This is the speed a body has
to reach to overcome the Sun's gravitational pull and
escape the Solar System. Any object entering the
Solar System for the first time would do so at its
own speed, plus the speed due to the Sun's gravitational pull, which is equal to its escape velocity. As
no meteorite has been observed with a speed relative
to the Sun that is greater than its escape velocity, and
since most meteorites are similar to the photographically recorded falls, it is reasonable to assume that
most meteorites come from within the Solar System
and are original residents of it.

The recent detection of particles entering the
Solar System from interstellar space has raised the

exciting possibility of capturing materials from
beyond the Solar System. Radar measurements suggest that the particles responsible for about 1 in 500
observed meteors could have orbits consistent with
an interstellar origin. Dust detectors on the Ulysses
spacecraft found a stream of particles coming from
the same direction as previously detected interstellar
gas. The findings were later confirmed by detectors
on the Galileo spacecraft, indicating about 12 parti-
cles/m²/day enter the Solar System from the direction of the star constellation Scorpio.

Radar meteor detectors in New Zealand confirm that interstellar particles regularly enter the
Earth's atmosphere. Precise measurements show that
some meteors have atmospheric speeds of more than
73 km/s (the maximum entry speed for objects in
'bound' orbits around the Sun). A smaller number
of these meteors have entry speeds of more than
100 km/s, the clear hallmark of an interstellar origin.
Larger objects in the Galaxy may periodically enter
the Solar System, but to what extent they may contribute to meteorites on Earth is currently a matter
for speculation.

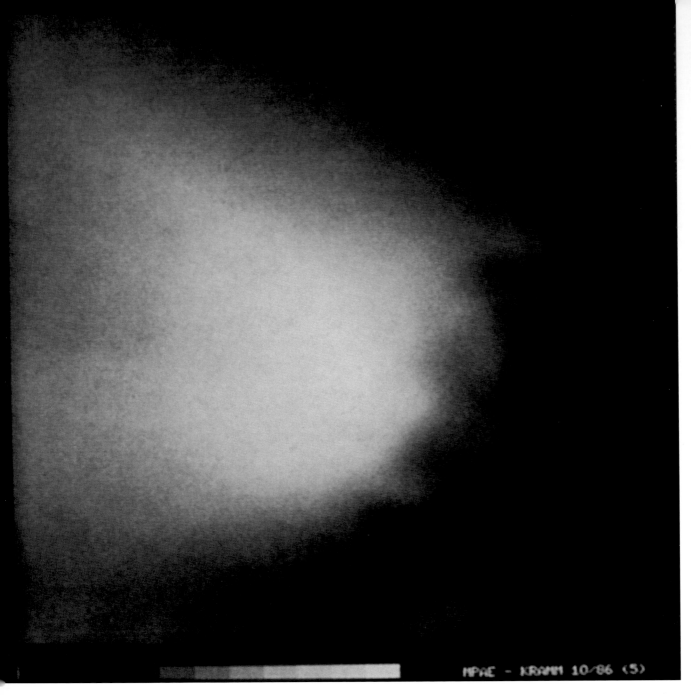

MPAE - KRAMM 10/86 (5)

Comets

Other potential, although unproven, sources of meteorites and micrometeorites are comets. These enigmatic nomads of the Solar System consist of loosely bound frozen water, ammonia, methane, carbon dioxide and millions of chunks of rock and grains of dust. Insignificantly small compared with the Sun and the planets, the core (or nucleus) of a comet might be up to 10 km across. But the combined mass of all comets is great, possibly about 50 times the mass of the Earth. The importance of these members of the Sun's family is that they, like meteorites, could represent samples from the early Solar System.

How comets formed is disputed. One suggestion is that during the birth of the Solar System, most of the lighter materials and gases were swept to the cold outer regions away from the Sun where they condensed to form a huge 'cloud' of dust particles and comets that now lies beyond the orbit of Pluto, and possibly extends half-way to the nearest star. This unseen cloud is called the Oort Cloud after the Dutch astronomer who first postulated its existence.

Perhaps containing millions of objects ranging from microscopic grains of dust to huge rocky icebergs tens of kilometres across, the outer part of the Oort Cloud cocoons the Solar System. A closer source of comets, the Edgeworth-Kuiper Belt, lies beyond the orbit of Neptune at a distance of around 4500-15 000 million km from the Sun. Both the part of the Oort Cloud closest to the Solar System and the Edgeworth-Kuiper Belt may be disk-like, orbiting around the Sun in the same direction as the planets. Although Pluto is a dwarf among the giant planets, it is actually a large member of the Kuiper

Viewed at a distance of 5000 km by the European Space Agency's Giotto spacecraft, the 10-km diameter nucleus of Comet Halley (left) erupts jets of gas from its dark surface.

belt objects that has come into a safe orbit at the edge of the Solar System.

When nearby stars pass close to the Oort Cloud, gravitational disturbances may cause comets to shoot off in random directions. Some fall towards the inner Solar System becoming trapped in orbits around the Sun. Such captured comets are called 'periodic', as they appear at regular intervals varying from less than a couple of hundred years to hundreds of thousands and even millions of years. Comet Halley, perhaps the most famous comet of all, completes a round trip through the Solar System every 76 years. Last clearly visible from Earth in 1986, Comet Halley will not pay us another visit until 2062.

Most periodic comets taking less than 200 years to complete one round trip through the Solar System move in the same sense as the planets around the Sun. Their orbits lie close to the plane of Earth's orbit indicating that their source is in the same plane. If these comets originated from the Edgeworth-Kuiper Belt then their orbits support the suggestion that it is disk shaped. Other comets enter the Solar System from all directions indicating that they came from the Oort Cloud, and many of these orbit the Sun in the opposite sense to the planets.

As a comet nears the Sun, heating causes frozen material to boil off its surface. Illuminated by sunlight, the liberated gas and dust forms a brightly glowing mantle, or coma, that may reach a diameter of over 100 000 km. The blast of radiation and atomic particles, called the 'solar wind', streaming out from the Sun forces the fluorescent gas and dust into tails. Consequently, the gas and dust tails of comets always point away from the Sun. As the comet returns to the outer reaches of the Solar System,

radiation from the Sun weakens and the comet dims.

After numerous orbits, each with a close encounter with the Sun, dust grains and small rocky fragments released from comets are smeared into broad streams in their wake. Periodically, the Earth crosses these streams of tiny particles and spectacular meteor showers sometimes result. Debris from Comet Halley causes the showers of meteors called the Aquarids and Orionids that regularly streak across our night skies in May and October respectively. The occasionally spectacular Leonid meteor shower in November is related to Comet Tempel-Tuttle, which orbits in the opposite sense to Earth, with its most distant point from the Sun lying beyond the orbit of Uranus.

Some meteor showers appear to have no source, but represent debris from periodic comets that disintegrated in the past. Importantly, meteorites do not fall more frequently during meteor showers. Most of the debris from comets consists of small particles usually weighing much less than 1 g. Because of their small size and excessive speed, this debris burns up high in the Earth's atmosphere, appearing as meteors. Analysis of the light emitted by bright meteors, however, suggests that cometary particles may have compositions close to certain rare types of carbon-rich meteorites, which themselves are similar to some dust particles. In addition, a few asteroids have comet-like orbits and may be the rocky cores of comets that have long since lost their ices and gases. These tantalising, but indirect lines of evidence suggest that some rare types of meteorite might come from comets.

When they are ejected from the Main Belt of asteroids, large meteoroids are attracted to the Sun,

Comet Halley was last visible to the naked eye from Earth in 1986.

The orbit of Comet Halley extends to outside that of Neptune, and takes 76 years to complete.

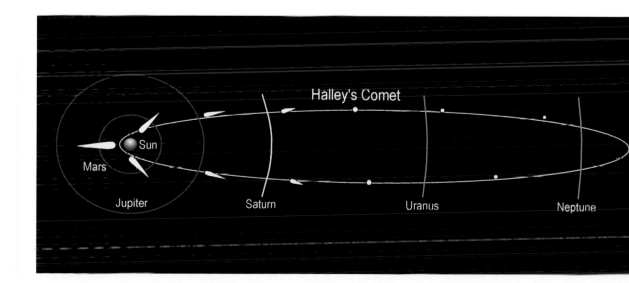

Halley's Comet

Sun

Mars

Jupiter

Saturn

Uranus

Neptune

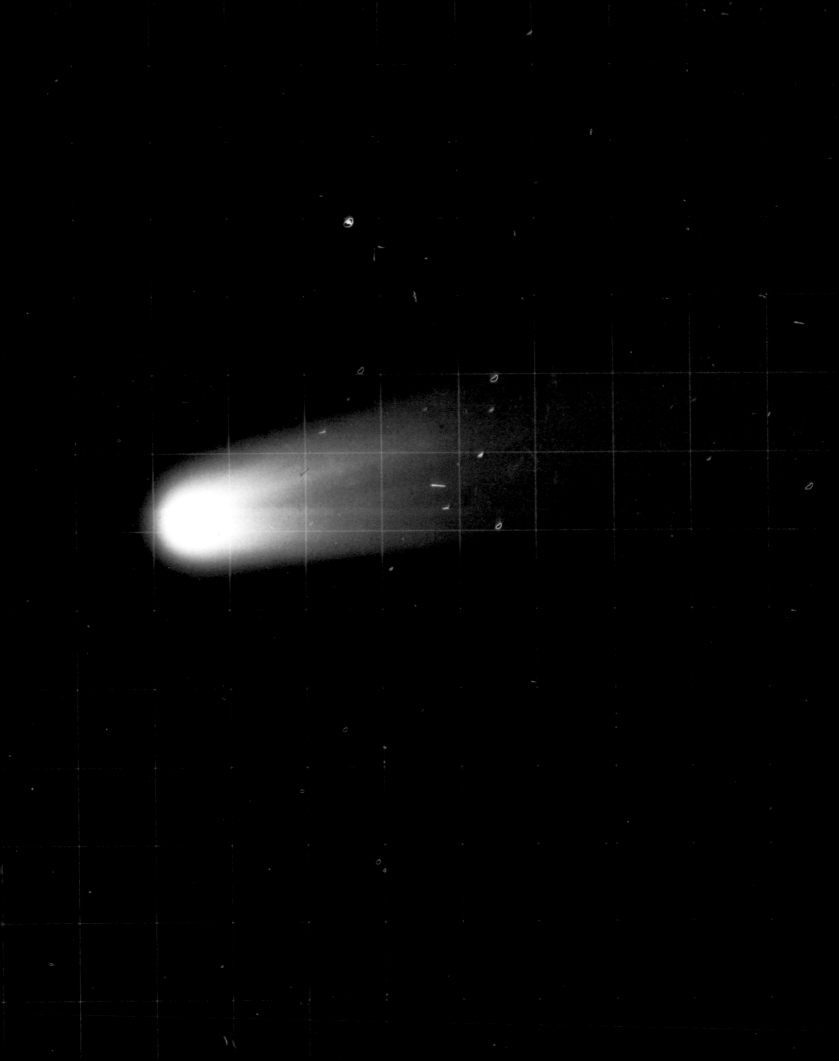

and their initial orbits are determined by the Sun's gravity. Their fate and changes in their orbits lie in eventual encounters with other bodies. Dust particles, however, behave differently from their larger cousins. The smallest particles are blown away by radiation pressure from the Sun.

At the mercy of the Sun's gravitational pull and the influence of large planets, larger dust particles also feel the effect of the solar wind. 'Drag' caused by a combination of gravity and the pressure of the solar wind changes the orbits of dust particles as they spiral slowly towards the Sun, eventually making them more circular. The faint glow seen before sunrise in the autumn, or the evening in spring, called the Zodiacal Light, is due to the scattering of light by a concentrated belt of dust extending from the Sun to beyond the orbit of the Earth.

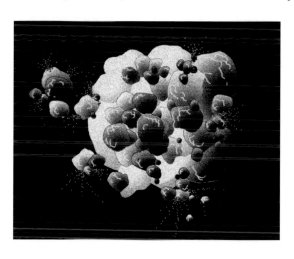

Possible sources of interplanetary dust particles and micrometeorites include both comets and asteroids, but their origins are still not fully understood. They represent important Solar System matter complementing meteorites, and their study over the last 25 years has given great insight to the nature of the early Solar System.

While a cometary origin of some meteorites remains unproven, recent discoveries have shown that meteorites are not exclusively samples of asteroids. Sixteen meteorites recovered from the Antarctic ice, one from Western Australia and two from the Sahara are fragments of the Moon. Another 21 are pieces of Mars. However, the majority of meteorites observed to fall on Earth are undoubtedly fragments of asteroids and, so far as we know, appear to have originated within the Solar System.

So, as the result of: Jupiter's gravity; collisions between asteroids; meteoroid impacts on moons and planets with low gravity and thin atmospheres; the slow destruction of comets; and the influx of grains from interstellar space, a large amount of debris is constantly released in the Solar System. The objects range in size from asteroids and comets to smaller fragments giving rise to recoverable meteorites and micrometeorites, with their sources varying from the inner to the extreme outer Solar System.

Above: The cometary nucleus is thought to be made of weakly bound fragments and dust. Periodic surface eruptions of gas and dust result from heating when near the Sun, and small orbital fragments may be bound gravitationally.

He handles the planets and weights their dust,

he mounts on the comet's car

And he lifts the veil of the sun,

and stares in the eyes of the uttermost star

'GRANDMOTHER'S TEACHING', ALFRED AUSTIN, 1888

SEARCHING FOR THE PAST

When Alfred Austin composed these lines of a poem more than a century ago, he could have been writing the job description of a meteoriticist. One frequently asked question is: why do you need more meteorites? Michael Lipschutz (Purdue University), likens deciphering the origin of the Solar System from meteorites to solving a jig-saw puzzle with less than 5 per cent of the pieces. The more meteorites there are for study, the better we are able to see the picture. Given that observed meteorite falls are extremely rare, this meagre material has to be supplemented from another source — meteorite finds, either by chance or searching.

One early pioneer of meteorite hunting was Harvey Harlow Nininger. This American biology teacher became fascinated by meteorites in the early 1920s. Touring the prairie states of the mid-west of the USA, Nininger lectured local groups and visited farms. By teaching people to recognise meteorites, he eventually gathered a large collection. The rolling plains of Kansas, Oklahoma, Iowa and Nebraska are generally rock-free and highly cultivated. Many unusually heavy rocks turned up by ploughs and tossed into stone piles turned out to be meteorites.

In 1972, there were only about 2100 meteorites known to science, just over half of which were 'finds'. This surprisingly low total represented a world-wide recovery rate of about 10-12 meteorites (falls and finds together) each year for the previous two centuries. When you leaf through the world

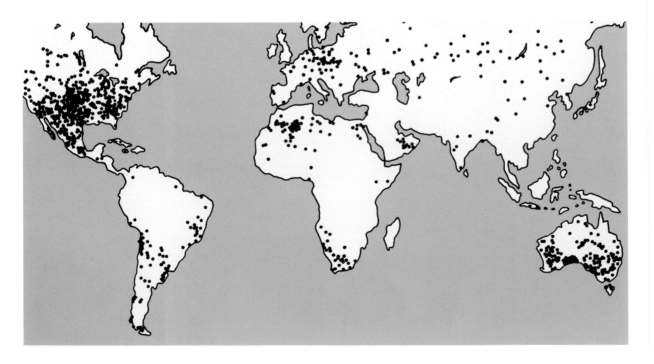

Meteorites have been found all over the world. The concentration of finds in several desert areas shows that those conditions have been favourable to their prolonged preservation and easy recognition.

listing of meteorites, the *Catalogue of Meteorites*, it reads like a Gazetteer. Far away places with strange sounding names adorn every one of its nearly 700 pages. With the frequency of American names, it's not long before you realise that up until fairly recently the USA was the country where most meteorites had been found. Over the last 30 years, serendipity combined with great human endeavour in some of the most remote and inhospitable places on Earth has yielded a bonanza of meteorites.

Survival of meteorites falling unobserved onto land depends mainly on climate. As Harvey Nininger showed, whether or not those surviving meteorites are found depends on our ability to recognise them. Meteorites are mixtures of minerals, some of which are unknown on Earth, including a few, like metallic iron, that corrode rapidly in the Earth's moist, oxygen-rich atmosphere. In wet climates meteorites are destroyed quickly by weathering, but in deserts they can be preserved for tens of thousands of years.

Two areas of the southern hemisphere, Antarctica and Australia, and two of the northern hemisphere, the deserts of North Africa, and the 'badlands' of Roosevelt County in New Mexico, USA, are prolific sources of meteorites. Namibia, in south-west Africa, also contains accumulated meteorites. A few other deserts, such as the Jiddat al Harasis in Oman in the Arabian Peninsula, have yet to show their full potential.

The first meteorite found in Antarctica was at Adelie Land, in 1912, about 32 km west of Cape Denison during the Australian Antarctic Expedition (1911–14). Further sporadic finds were made by international expeditioners from Russia (two fragments of an iron meteorite at Lazarev) and the USA (two fragments of a stony iron at Thiel Mountains)

in 1961. Another small mass of meteoritic iron was found by Americans in 1964 at Neptune Mountains.

Five months after astronauts stepped on to the Moon for the first time, in December 1969 geologists from the 10th Japanese Antarctic Research Expedition were walking across 'blue' ice near the Yamato Mountains when, in a short time, they found nine fragments of meteorites. Not until much later was the significance of this discovery appreciated. Eventually shown to include several different types of meteorite, the recovered fragments must have come from different falls.

Where the Antarctic ice-cap piles up behind a barrier, like a mountain, fierce Antarctic winds strip it away revealing compressed 'blue' ice. After the finds at Yamato came the realisation that meteorites that had fallen at different times were entombed together in the ice. Ice movements concentrate the cargo of meteorites in areas of stagnant flow, where wind erosion exposes them at the surface. A Japanese search party returned to the Yamato Mountains in 1973 and found another 12 meteorites, but in the following year searchers recovered a staggering 663 meteorites. Since 1973, repeated expeditions to the Yamato Mountains and other areas of Antarctica have recovered more than 8500 fragments of meteorites.

In 1976, joint teams from the USA and Japan began searching for meteorites in Victoria Land, on the opposite side of Antarctica to the Yamato Mountains. New areas of blue ice yielding meteorites were found near the Trans Antarctic Mountains, proving that discoveries at the Yamato Mountains were not unique. To date, more than 20 000 meteorites have been recovered from more than 40 sites in Antarctica. With great success, groups from other countries, including Italy, France, Germany and the

United Kingdom have searched areas of Antarctica for meteorites. Many other promising sites have yet to be explored. A few Antarctic meteorites have resided in the ice for more than a million years.

The separate meteorite falls represented by the wealth of material from Antarctica are difficult to estimate. Many fragments belong to showers of common meteorite types, not easily distinguished one from another. Conservative estimates place the number of different falls at around a tenth of the total number of fragments recovered. So, intensive collecting in Antarctica has more than doubled the number of meteorites known to science.

Following the Antarctic discoveries came a gradual realisation that some deserts also contain abundant meteorites. Australia is high on the list of those areas of the world with large tracts of arid land. Many thousands of pieces from more than 500 or so distinct meteorites, 97 per cent of which are finds, have been recovered in Australia. Resulting from an unique environment, abundant meteorite finds from the Nullarbor Plain in Australia account for more than half the total number so far recovered from the island continent.

At the heart of a much larger, but geologically related area lies the Nullarbor Plain. Representing a time when the Great Australian Bight bit more deeply, the flat-lying, 13-million-year-old limestones of the Nullarbor now form a low-lying plateau.

Left: Tiny meteorites (small black speck) are easily spotted on the Antarctic ice.

Below: Search parties in Antarctica 'flag' meteorites as they are found.

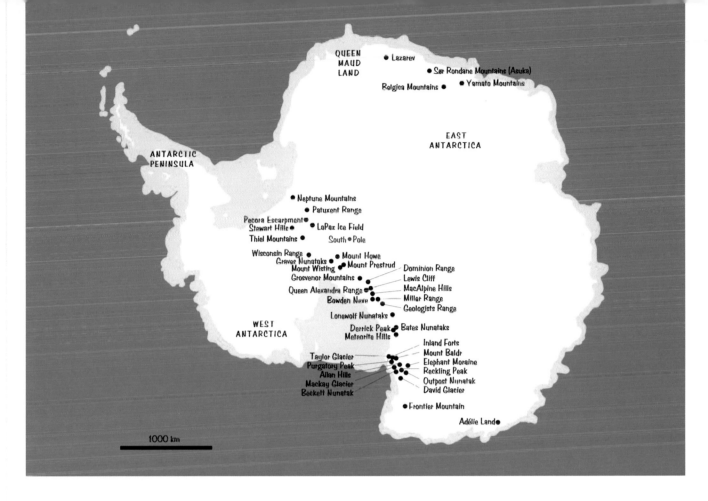

More than 20 000 meteorites have been collected from 40 localities in Antarctica.

At 5.5 kg, the Yamato-74077 stony meteorite was the largest found by the Japanese party during the 1974-75 field season in Antarctica.

One of the largest, featureless areas of the Earth's surface, this vast plain covers an area the size of Oregon. Its name, from Latin *nulla arbor* meaning 'without trees', reflects a climate parched for tens of thousands of years. Bone-dry conditions have preserved ancient meteorite falls on the Nullarbor, while the lack of vegetation is ideal for meteorite hunting. But this stony desert has another advantage. Because meteorites are generally dark, or rusty-looking rocks, they stand out against the pale limestone littering the plateau.

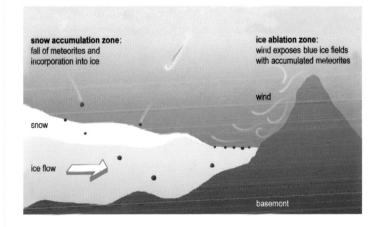

Meteorites falling over time are trapped in the Antarctic ice, and transported and concentrated in areas in front of mountain barriers, eventually to be brought to the surface again in 'blue' ice areas revealed by wind erosion.

ALL IN A DAY'S WORK

Sooner or later it had to happen, but the telephone call in July, 1988 still took me by surprise.

'Alex is that you?', asked the unmistakable voice of John Carlisle. 'You'd better come out here. I've found a big iron meteorite — two tons I reckon.'

'How will we get it back', I ask barely able to contain my excitement. 'Rod Campbell says he'll do the job — see you at Kybo Station', came the reply.

By chance, Bob Hutchison, then Curator of Meteorites at the Natural History Museum in London, and his wife Marie were soon to arrive from England to join me and Tom Smith (Perth Observatory), on one of our regular meteorite collecting expeditions to the Nullarbor.

A couple of weeks later we are in Kalgoorlie. From there we take the track running east beside the Trans-Australian Railway. Five hot and dusty hours down the tree-lined track, the eucalypts suddenly give way to open scrub. Soon there are only blue-bush flats stretching before us, the bare, brown clay dotted with lumps of white limestone. We are on the Nullarbor Plain.

At Rawlinna, we stop for a rest. The shop at the station was then the last until you got to Cook in South Australia, more than 500 km to the east. Refreshed, we press on. Along the winding, boulder-strewn track it takes us four bone-shaking hours to cover the next 130 km. At dusk we arrive at Kybo Station, and Rod and Jill Campbell are there to

greet us. But the journey is not over yet. John Carlisle, Greg and Colin Campbell have taken the truck on ahead. Burdened with a tractor, the truck can barely travel more than 10 km/h. Soon we catch up with them, their huge campfire of discarded railway sleepers shining like a beacon across the rolling plain.

By noon on the following day our convoy arrives at the place where the 11.5- and 6.1-ton Mundrabilla meteorites were found in 1966. A concrete plinth and brass plaque proudly marks the site. We are pretty sure that John's new discovery is another chunk of the same unusual iron. A few kilometres further we climb over a small rise and there is the meteorite. About 1.5 m across, its grey, knobbly surface thrown into sharp relief by the lengthening afternoon shadows. Decorated with cobwebs and surrounded by soil trampled by foxes and rabbits, the meteorite has been a home and a playground for generations of creatures.

'That's not two tons', says Bob Hutchison using his experienced eye. 'More like four!'

He is soon proven right. Rod's boys dig away at the base of the meteorite securing it in a cradle of chains. Just as Bob predicted, we are looking at the tip of the 'iceberg'. A third of the meteorite is buried in the soft clay. Rod's innovative plan is to use the shovel on the tractor to hoist the meteorite on to the back of the truck in piggy-back fashion. Giving it all

The 3.5-ton fragment of the Mundrabilla meteorite was recovered from the Nullarbor in 1988.

Prolonged dry conditions and a stable surface preserve meteorites in desert areas like Australia's Nullarbor Plain where they have accumulated over the last 35 000 years. Mostly dark rocks, meteorites are easily spotted against the pale limestone of the plain.

he can, Greg Campbell works the controls, but the fountain of oil from a ruptured hydraulic line says it can't be done.

Leaving the recovery job to the 'professionals', we slip away to collect the hundreds of small fragments of iron littering the ground around the big meteorite. Quickly repairing the hydraulics, Rod decides to revert to classical methods of meteorite recovery. Building a ramp to get the tractor off the back of the truck, the shovel is used to dig a hole next to the meteorite. We return in time to see Rod back his one-ton pick-up into the hole.

'He's not going to load it on to that, is he?', Bob exclaims.

'That's exactly what he's going to do', I reply trying to disguise my own misgivings.

Skilfully manoeuvring the shovel, Greg Campbell gently nudges the meteorite onto the tray of the pick-up. With a deafening screech of cosmic

metal on man-made steel, and the creaks and groans of incipient fatigue, the meteorite slides into place.

The next problem: How do you get a pick-up carrying a 3- or 4-ton meteorite out of a hole? You pull it out with a tractor.

With inverted springs, visibly wilting under the strain of a load that its designers would not have considered in their wildest nightmares, the pick-up looks distinctly unwell. Strengthening the tray with a couple of hefty iron bars, Rod administers first aid to the ailing vehicle. There are now only 3 cm between the wheel arches and the tyres. All that remains is to deliver the heavenly body to the railhead at Loongana about 60 km across country — at night!

In the gathering gloom, the Campbells set off. Soon the whine of the labouring pick-up and the roar of the truck merge with the howl of the prevailing south-westerly Nullarbor wind. The rest of the party stays behind to collect meteorites.

Fourteen hours and 17 punctured tyres later (miraculously only on the truck), the Campbells pull in to the sleepy township of Loongana. A freight wagon waits on the line to take the precious cargo to Perth. The tractor shoves the meteorite into the wagon. Up goes the pick-up, down goes the wagon and the job is done.

'We can move mountains out here', says Rod.

If there were any on the Nullarbor, I would believe him.

Right: The Libyan Sahara is another region where meteorites, such as this ancient weathered meteorite from Sarir al Qattusah (Dar al Gani), have been recovered.

Left: The main 11.5-ton mass of the Mundrabilla meteorite was transported from the Nullarbor to the Western Australian Museum in Perth in 1966.

The first meteorite found on the Nullarbor was in 1911 during the construction of the Trans-Australian Railway, but it was not until the mid 1960s that its true potential as a source of meteorites was realised. Many successful forays were made into the desert to search. During the late 1960s and early 1970s the number of meteorites known from Australia increased by 57, with the biggest contribution coming from the Western Australian Nullarbor.

Not including the 11.5-ton Mundrabilla meteorite, the largest ever found in Australia, during the 1960s alone nearly 1 ton of other meteorites passed into collections. The haul included 37 new meteorites, adding about 2 per cent to the total number known in the world at that time. Many of the earliest discoveries in the Nullarbor were made by rabbit trappers, but the largest single contribution was made by John Carlisle, a professional bushman from Kalgoorlie, and members of his family.

As the result of intensive searching since the mid-1980s, the Nullarbor has given up several thousand fragments from more than 300 different meteorites. Among this wealth of material there are many hundreds of fragments of potentially new meteorites yet to be studied. Joint teams from the Western Australian Museum and EUROMET (a pan-European group of research institutions devoted to meteorite research) found more than 600 meteorites during only ten weeks of searching in the Nullarbor on four expeditions between 1992 and 1994.

While the Australian crop of meteorites results from a prolonged dry climate, in Roosevelt County, New Mexico, other reasons contribute to the cosmic harvest. There meteorites accumulated with time in dry, wind-borne sediment called loëss. Over tens of thousands of years this fragile, powdery soil was winnowed by strong winds, alternately burying and exposing meteorites. Today, where soil has blown away, the cargo of meteorites now rests on stable stranding surfaces.

On the Nullarbor Plain, five people standing 10 m apart can search 1 km² every 20 km walked.

orites have been found in the wider Sahara including parts of Mauritania, Niger, Tunisia and Egypt.

Not all deserts are littered with meteorites. South American deserts like the Atacama in Chile contain meteorites, but they have not been recovered in large numbers. Flanked by the Andes Mountains, the Atacama is strewn with dark volcanic rocks that hamper meteorite recognition, while the desert's shifting surface often buries small meteorites.

Deserts in China and Mongolia, such as the Gobi, seem good prospecting grounds for meteorites but few have been found. Adjacent to mountains periodically generating rain, these deserts are now dry but they were much wetter in the past. Torrential downpours scoured their surfaces burying meteorites or rendering them unrecognisable. In the young deserts of China, with surfaces around 5000 years old, there has not been sufficient time to concentrate meteorites. China, however, is covered in vast deposits of loess. The possibility that meteorites have accumulated in Chinese loëss, like in Roosevelt County, has yet to be explored.

The deserts of Inner Mongolia in northern China have geologically young surfaces, less than 5000 years old, and there has not been sufficient time for meteorites to accumulate in great numbers. Extensive searching by teams from the Chinese Institute of Geochemistry have been unsuccessful in finding meteorites.

Searching the deserts of north Africa, particularly the Libyan and Algerian Sahara, has revealed meteorite accumulations rivalling that of Antarctica. These stony deserts, or regs, are paved in pale country rocks with sandy surfaces. During the early 1990s, 471 meteorites were collected in Algeria and Libya. Another 65 meteorites had previously been found in other areas of Libya. Many meteorites were found in the Acfer-Aguemour region of Algeria, a small patch of desert measuring 30 by 100 km. Other important sites are the Daraj, Hammadah al Hamra, and Sarir al Qattusah (Dar al Gani) areas of neighbouring Libya. To date, more than 1000 mete-

WHAT'S IN A NAME?

Meteorite names normally carry no scientific information. They are simply geographical labels distinguishing one meteorite from another. Meteorites take the name of a permanent geographical feature closest to the site of their recovery. Far more numerous micrometeorites are not named, but numbered and grouped for study.

Charged with the task of approving meteorite names is the Nomenclature Committee of the Meteoritical Society. Founded in 1933, the society is devoted to all aspects of meteorite research. Names, and details of new falls and finds are published periodically in the *Meteoritical Bulletin*, a section of the society's journal *Meteoritics and Planetary Science*. Eventually, approved names are added to new editions of the *Catalogue of Meteorites* published by the Natural History Museum in London.

New editions of the *Catalogue of Meteorites* appear every 10-15 years. Inevitably, there is a long time between the fall or discovery of a meteorite and addition of information into a new edition of the catalogue. Up-to-date information on Antarctic meteorites collected by teams from the USA can be obtained from the *Antarctic Meteorite Newsletter*, a periodical issued through NASA Johnson Space Centre in Houston, Texas. Japanese Antarctic discoveries are documented in the *Photographic Catalogue of The Antarctic Meteorites* and *Meteorite News*, both published by the National Institute of Polar Research in Tokyo.

Although the geographical naming system works well in most parts of the world, it could not be used for the great numbers of meteorites from remote parts of the world. In Antarctica and the world's hot deserts, there are simply not enough names to go around. Antarctic meteorites take the name of the locality in which they are found, such as Yamato or Allan Hills, followed by the year in which they were found, and three digits signifying the specimen number. So Allan Hills 84006 and Yamato 84010 show that these were the sixth and tenth meteorites numbered from those localities, respectively, in the Austral summer of 1984-85. Shortening the names, Allan Hills becomes 'ALH' and Yamato 'Y'. Early names at the Allan Hills locality also recognised the field party that found a particular meteorite. For example, the ALHA 77005 meteorite was the fifth meteorite found by field party A in 1977.

To solve the naming problem in hot deserts, meteorites are tagged with a general locality followed by a three-digit number denoting the find, such as Roosevelt County 010, Dar al Gani 262, and Loongana 001. Because names are more memorable and charming than numbers, many 'meteoritophiles' mourned the demise of the old system. In the Nullarbor, exotic and often invented names such as Dingo Pup Donga, Mulga (west), and Laundry (east) are reminders of a more romantic era.

When viewed under a microscope, the glacial sand recovered from Cap Prudhomme on 14 January 1994 reveals irregularly shaped unmelted micrometeorites (dark particles), and some micrometeorites melted by atmospheric friction (spherules).

The great Armanty iron meteorite, weighing around 20 tons, is one of the tourist attractions of the city of Urumchi in north-west China.

Three steam generators of the French 'micrometeorite factory' at Cap Prudhomme in Antarctica inject jets of melt-water into drill holes to a depth of 5 m. Immersion pumps recycle the water to feed the steam generators, and after 8 hours of cyclic operation three pockets of water 3–5 m³ in volume are produced. Glacial sand deposited in the bottom of the pockets is drawn up by the pump and passed through stainless steel sieves. In a typical day's collecting, several thousand micrometeorites can be recovered.

Only around ten large collections, such as that of small meteorites in the Natural History Museum in London, provide most of the material for meteorite research.

Cruising at 20 km above the Earth, planes flown for NASA capture interplanetary dust particles as they fall to Earth.

Gathering dust

Searching for micrometeorites presents special problems, not least of which is their size. Cosmic spherules are usually less than 1 mm across, and interplanetary dust particles less than 0.1 mm. Although the infall of dust is large, it is quickly lost on the rocky surface of the planet, or settles in sediments in oceans and lakes. Several ingenious ways are used to collect dust. Catching particles before they fall to Earth is the best.

Since the late 1970s, high-flying aircraft from the United States' National Aeronautics and Space Administration (NASA) have collected dust from the Earth's atmosphere. Cruising at a height of around 20 km, aeroplanes extend special dust-catching panels coated with silicone grease. Withdrawing the panels before the plane descends prevents contamination from Earth dust. In ultra-clean laboratories the grease is dissolved, leaving the tiny particles on filters. More than 10 000 particles have been recovered in this way, but with a total weight of less than 1 g!

Cosmic spherules are recovered by dredging clay from the ocean floors, melting large quantities of ice in Antarctica and Greenland, and from a deep well supplying water to the Scott–Amundsen Station in Antarctica. Like cosmic spherules, interplanetary dust particles have also been recovered from the Antarctic ice. Magnetic cosmic spherules can also be collected by combing ancient lake sediments with powerful magnetic rakes.

The working sample

Of thousands of tons of extra-terrestrial material gathered from observed meteorite falls and by sustained collecting from deserts, there are more than 4 kg of lunar meteorites to supplement the 381.7 kg of Moon rocks returned to Earth by American and Russian space missions, and more than 70 kg of Martian meteorites. The great bulk of the material, however, originated from the asteroids.

Matching meteorites and asteroids, a subject to which we will return later, is a difficult but worthwhile exercise that has met with limited success. In collections there are probably around 300 kg of 4 Vesta, and possibly a large number of fragments of 6 Hebe. With a great deal of uncertainty, other meteorites have been linked to particular asteroids. This, then, is the sample with which we have to work. What meteorites and micrometeorites tell us about the history of the Solar System, and the bodies of which they were once part, is the subject of the next few chapters.

... like trying to learn about

flour and eggs by studying a cake.

RT DODD, 1986

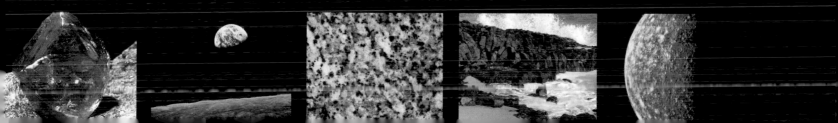

ANATOMY OF A PLANET

In a sense, cooking is only a specialised branch of chemistry — and a tricky one at that. Even the most experienced cooks suffer occasional failures. In baking, ingredients are added in certain proportions, carefully blended to the desired consistency, then heated. The result, be it a cake, a loaf of bread, or an exotic pastry, is very different from, and a great deal more appetising than, the starting materials. What's more, the change is irreversible. Try as you might you could not separate all those ingredients again. Most of the water in the original mix was driven off to waft around the kitchen as vapour, eventually condensing on the cold walls. Apply too much heat, then from the same ingredients there is a different result. Long after the charred remains have been consigned to the bin, the acrid smell of smoke lingers as a reminder of the failure. Nature's chemistry has no failures, only results. But just as in the kitchen, there can be more than one result from the same ingredients.

Our 'cake' is the Earth, and our interest lies in the original ingredients from which it was made. Since it was formed, our planet has changed constantly. Unlike for a cake, we do not know what Earth's original ingredients were. In the rocks seen at the surface today no evidence remains of Earth's assembly. But by studying the chemistry and texture of Earth's rocks, and gaining knowledge of its deep interior, a great deal can be learned about the events that moulded the planet. This also gives some insights into how rocks from other parts of the Solar System might have formed.

All rocks, including meteorites, are mixtures of minerals. Individual mineral grains are made of atoms of chemical elements in various regular, ordered, crystalline arrangements. Atoms are the building blocks of minerals that are, in turn, the building blocks of rocks. Minerals and rocks are made by essentially the same process, and rocks are recipes in the Solar System's historical cookbook. Written in a language only partly understood, this 'book' is not an easy read. Deciphering the cryptic language of rocks and minerals is the realm of geology and geochemistry.

In the 1920s, VM Goldschmidt, after studying the way elements distribute themselves between minerals in meteorites and Earth rocks, recognised that affinities existed between groups of chemical elements. Metallic elements, such as gold, almost always occur as metal, either on their own or alloyed with other metals. After the Greek *sideros* (iron) he called such elements 'siderophile', meaning metal-loving. Those elements on Earth with an affinity for sulfur, such as lead, Goldschmidt called 'chalcophile' (*chalkos* means ore). Elements such as silicon, magnesium and calcium that are oxygen-loving, commonly forming oxides and silicates, he called 'lithophile' (*lithos* means stone). While gaseous elements such as xenon, neon and nitrogen he called 'atmophile' (*atmos* is air).

Goldschmidt's description of element behaviour explains the main chemical components of the Earth, forming the basis for the modern branch of geology called 'geochemistry'. Under certain conditions, some elements behave in different ways. In the Earth, iron is siderophile, chalcophile and lithophile, distributing itself between metal, sulfides and silicates according to availability and conditions. Oxygen, amongst the most abundant elements on Earth is both atmophile and lithophile.

Today, Goldschmidt's original concept is extended to include planetary science and cosmochemistry. Because of differences in the overall chemistry and situations of the planets and other bodies in the Solar

Earth rise over the Moon. Of the planets in the Solar System the Earth is the most geologically active. Since Earth's formation around 4 500 million years ago it has changed constantly. Today no evidence of the planet's original assembly is to be found at the surface, yet estimates of the composition of the whole Earth play a part in our understanding of the evolution of the Solar System.

ELEMENTS AND ISOTOPES

Everything is made of atoms of chemical elements. There are 115 elements known, but only 91 occur naturally in detectable amounts. Chemical reactions affect the binding of atoms to make molecules, but not the nature of the atoms themselves. The identities of atoms are changed only by nuclear reactions. The life-giving light and warmth of our Sun results from energy released by nuclear reactions.

Hydrogen (with the symbol H) is the lightest and most abundant element in the Universe, making up 93 per cent of the total number of atoms or 76 per cent of the mass. Helium (He), the next most abundant, makes up 7 per cent of the atoms and 23 per cent of the mass. All the elements heavier than helium together amount to little more than 1 per cent of the mass of the Universe. Hydrogen and helium together make up 99 per cent of the weight of the Sun. Atoms can be imagined as tiny electrons (negatively charged particles) around a small central nucleus. In the nucleus there are protons (positively charged particles) and neutrons (carrying no electrical charge) which together give the atom its weight. The number of protons in the nucleus of an atom distinguishes one element from another. For example, most hydrogen atoms consist of one proton and one electron, and have an atomic weight of 1. A small proportion of hydrogen atoms have one neutron as well. Deuterium (D) is an isotope (pronounced *ice-o-tope*) of hydrogen. Isotopes are atoms of the same element containing different numbers of neutrons. This 'heavy' hydrogen, called deuterium, has an atomic weight of 2. Only about one in every 5000 molecules of ordinary water (hydrogen oxide or H_2O) on Earth is made of deuterium oxide (D_2O). A third, even rarer, isotope of hydrogen (tritium) has two neutrons and an atomic weight of 3.

Some isotopes are stable, that is to say they are unchanging. Unstable, radioactive isotopes change, or decay, to other isotopes liberating energy in the form of heat. The rate at which this nuclear decay occurs is expressed in the amount of time taken for half the total number of atoms of a radioactive isotope to decay. Decay rates, or half-lives, of a number of radioactive isotopes are known. Hydrogen and deuterium are stable isotopes, but tritium is unstable and decays with a half-life of 12.3 years. Since the discovery of radioactivity in 1896, knowledge of this property has provided a precise method for dating of rocks.

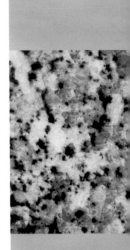

Above: Granite makes up most of the Earth's continental crust. It is a mixture of three main minerals: quartz and feldspar (light-coloured crystals), and biotite mica (dark).

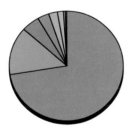

- silicon dioxide 72.5%
- alumina 14.1%
- potash 5.2%
- soda 3.3%
- iron oxide 2.5%
- calcium oxide 1.6%
- magnesium oxide 0.5%
- titanium dioxide 0.3%

In alumina, two aluminium atoms combine with three oxygens; in potash, two potassium with one oxygen; in soda, two sodium with one oxygen; silicon and titanium atoms combine with two oxygens each; and calcium, iron and magnesium combine with one oxygen each. These eight compounds are the main chemical components of granite (above) and basalt (below).

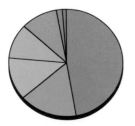

- silicon dioxide 47.3%
- alumina 16.3%
- calcium oxide 12.8%
- iron oxide 10.5%
- magnesium oxide 10.2%
- soda 1.8%
- titanium dioxide 1%
- potash 0.1%

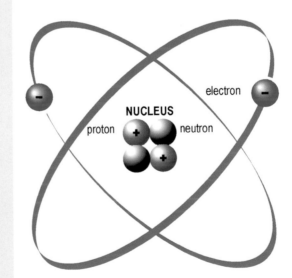

A helium atom comprises a nucleus of two protons and two neutrons, around which are two negatively charged electrons. Helium has an atomic weight of 4, and is the second-most abundant element in the Universe, after hydrogen.

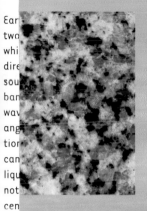

Below: At Bunbury on the west coast of Western Australia, a solidified lava flow, which erupted when India broke away from Australia around 130 million years ago, shows typical columnar jointing.

System, the behaviour of elements can change with prevailing conditions. Venus is an inhospitable planet with a surface temperature of around 400°C. On Venus sulfur is partly atmophile as sulfur dioxide. A typical Venutian weather forecast might predict intermittent sulfuric acid showers!

To make sense of a large number of materials, scientists put them into some sort of order. The first step is to group similar materials. In the same way as we seek our ancestors, Earth scientists seek relationships between different 'families' of materials. They have at their disposal an impressive array of powerful analytical tools allowing rocks and minerals to be grouped together on much more than their outward appearance. They also visit places to collect rocks, observing the relationships between them at the sites where they occur. As it is the most geologically diverse body in the Solar System, Earth exhibits abundantly the processes that make rocks.

On Earth there are three main kinds of rocks; igneous, sedimentary, and metamorphic. These group names tell us how the rocks formed, but not what they are made of. Igneous rocks (from Latin *ignis* meaning fire) form by the solidification of molten rock, or magma, generated deep within the Earth. Usually, when magma cools, crystals of minerals grow until the liquid is consumed. If magma is erupted at the surface as lava, rapid cooling leaves little time for crystals to grow so they are small. In some cases cooling is so rapid that lava solidifies as a glass. If magma is trapped at depth and cools slowly, large crystals can form.

Granite is an igneous rock made of visible crystals of the minerals quartz (silicon dioxide), sodium- and potassium-rich feldspars and usually mica, which all grew by slow cooling deep within the Earth. By contrast, chemically identical rhyolite, a rock in which it is difficult to see crystals with the naked eye, is erupted at the surface of the Earth through volcanoes. Gabbro and its volcanic equivalent, basalt, are igneous rocks usually made of the minerals olivine (iron-magnesium silicate), pyroxene (calcium-iron-magnesium silicate) and calcium-rich feldspar.

basalts are common on those planets as well. Basalt volcanism, then, is a characteristic of planetary evolution.

During melting and separation of planetary
objects two main processes — fractional crystallisation and partial melting — operate to produce a
variety of igneous rocks. When the first formed crystals separate from magma and do not change the
remaining melt by reacting with it, the process is
called fractional crystallisation. One way for crystals
to separate and become isolated is by sinking to the
bottom of the melt to accumulate as a crystal mush.
The rocks so formed may be layered and are called
cumulates. The remaining melt has less of the elements that made the early crystals. Because it crystallises at high temperatures, olivine is often the first
mineral to form. Olivine removes more magnesium
from the melt than iron, and also leaves it richer in
silicon. Crystals of other minerals growing later than
olivine may reflect this melt composition and have
lower amounts of magnesium compared with iron.
Partial melting is essentially the opposite of fractional
crystallisation. The first liquid to form during partial
melting has the same composition as the last liquid
to crystallise during fractional crystallisation.

One recipe

Obtaining a composition for the whole Earth is not
an easy task, for there are regions of its interior about
which we know very little. Although we do not
know exactly the range of primitive materials that
formed the Earth, or indeed the relative proportions
in which they were mixed, from the gross features of
the entire planet we can calculate a rough overall
chemistry. One way to obtain part of the chemistry
of the Earth is to look at rocks brought to the surface from deep within the planet by volcanoes.

Throughout the continents and oceans of the
world, basaltic lavas occasionally contain bright-green
lumps of granular, olivine-rich rock. Called olivine
nodules, the jewelled appearance of these foreign
bodies contrasts starkly with the black, frothy lava
that delivered them from depth. These nodules from
the deep also contain emerald-green crystals of
chromium-bearing pyroxene, and spinel, a dense
oxygen-bearing mineral containing aluminium, magnesium, iron and chromium.

Minor elements tightly bound in the atomic
structures of minerals in olivine nodules reveal that
their mineral make-up is stable at high temperatures
(up to 1400°C) and extreme pressures up to 20 000
times greater than at the Earth's surface. Such hot,

pressurised conditions should exist at depths of
around 60 km below the Earth's surface. Olivine
nodules must then come from deep within the
Earth, perhaps from the upper part of the mantle.
Because nodules were brought swiftly to the surface
their minerals retain their high-pressure state.

Several other rocks have a deep Earth origin.
Especially interesting, because of their economic
value, are rocks containing diamonds, like kimberlite.
Diamonds can be made in the laboratory at 30 000
times atmospheric pressure and temperatures over
700°C. Naturally, these conditions are found only at
great depth in the Earth's mantle, about 150-200 km
down. Under such high pressure and temperatures,
carbon crystallises to form diamond, not the other
common mineral form of pure carbon, graphite.

Only rapid transport in volcanic pipes, at about
70 km/h, from the mantle to the Earth's surface
saved diamonds from being destroyed or converted
to graphite. Most common magmas, such as basalt,
form in the upper 50 km of the Earth's crust. More
unusual magmas, such as kimberlites and lamproites,
originate at 150-200 km depth. These rocks are like
basalts but contain more magnesium and potassium.
Jetted to the surface by violent volcanic gas explo-
sions, a range of high-pressure rocks and minerals are
preserved in kimberlite because rapid expansion of
gas led to their cool, low-pressure emplacement.
Eclogite and garnet-peridotite are two other rocks
that formed under high pressures brought to the surface from great depth by kimberlite eruptions.

Dominated by olivine, garnet-peridotite lacks
spinel, and most of its aluminium is stored in a red,
magnesium-rich variety of the mineral garnet.
Garnet with this composition is typical of rocks
formed under high pressures, and is a good pointer
to the possible presence of diamonds. Eclogite, a
beautiful rock made of deep red garnet and green

Olivine nodules, made
mainly of olivine (pale
green) and chromium-
bearing pyroxene (dark
green), are brought to
the surface from depths
of up to 60 km encased
in volcanic basalt lava
(black). Such nodules
provide important
information about the
composition of the
upper mantle. (View
about 10 cm across.)

- iron 33%
- oxygen 29.8%
- silicon 16.9%
- magnesium 14.6%
- nickel 1.7%
- calcium 1.6%
- sulfur 1.5%
- aluminium 0.8%
- sodium 0.1%
- titanium 0.1%

- iron 89.9%
- nickel 5.4%
- silicon or oxygen 4.7%

The composition of the whole Earth (top), and its core (bottom).

pyroxene, results from extreme pressure-induced development of minerals with high densities. Squeezed to their limits, the rock's minerals change to ones occupying less space.

Olivine nodules, eclogite and garnet-peridotite are several different samples of the mantle, their analysis revealing its approximate composition. But this is an estimate of today's mantle, not of the Earth's 'primordial' mantle. The loss of basalt lava erupted at the surface over time has changed the composition of the upper mantle. When 'mantle' rocks such as eclogite are partly melted in the laboratory they first yield around 10 per cent of a basalt-like liquid. Most of the aluminium, calcium and sodium goes into this liquid, leaving the remainder impoverished in these elements but richer in magnesium. Allowing for the loss of this 10 per cent of basalt gives an indication of the chemistry of the early mantle.

Armed with an estimate of the composition of the mantle, a rough composition for the whole Earth can be calculated from its major components. The Earth's inner and outer core make up 32 per cent of the weight of the planet. When the densities of these regions are adjusted to surface pressure, the core has a density slightly less than pure iron, with the outer core a little less again. Accounting for the discrepancy, around 20 per cent of the outer core, less of the inner core must be made of material with a density less than iron. By analogy with meteorites the total core probably contains 28 per cent of the Earth's iron, but also contains nickel and some lighter elements, including sulfur.

Ignoring the crust, oceans and atmosphere, the mantle makes 68 per cent of the Earth. Blending 68 per cent mantle with 32 per cent core we arrive at an estimate of the Earth's composition. When iron bound to oxygen in the mantle is added to free metallic iron in the core, the Earth contains by weight around 33–35 per cent iron. This kind of calculation gives reasonable estimates for only the most abundant chemical elements in the Earth. Later we will compare this rudimentary composition of the Earth with meteorites. But what about its age?

The age of the Earth forms the base-line for many branches of science, including geology and biology. In the history of science, obtaining an accurate age for the Earth was a great step in understanding both the planet itself and the Solar System as a whole. Because it has changed so much, the ages of rocks now exposed at the Earth's surface are bound to be less than the age of its assembly. In fact, the age of the Earth cannot be determined directly, but has to be constrained from a study of meteorites.

An ancient Earth

With our average life-span of about 70 years, periods of millions (let alone thousands of millions) of years are hard to imagine. By converting time into distance, we can gain a perspective of the great time spans involved. If the age of the Solar System and Earth were represented by a journey across Australia from Sydney to Perth (about 4000 km), each kilometre would be just over 1 million years. The Sun, planets and asteroids formed before we left the front of the Sydney Opera House. In Melbourne, 1000 km from Sydney, we would find the first primitive single-celled organisms. On the Nullarbor, 3000 km from Sydney, we would see our first plant life. At 600 km from Perth, Kalgoorlie would be inhabited by primitive, jellyfish like creatures. Cunderdin, only 150 km from Perth, would be overrun by dinosaurs. Even with only a short distance to go, we would not see anything looking remotely like a human being until we were walking down St George's Terrace in the centre of Perth.

Diamond crystals, such as those from the Argyle diamond mine in the Kimberley of Western Australia, are formed 150–200 km beneath the Earth's surface. (Approx 8 mm across.)

How do we know that meteorites and the Earth are so ancient? The answer lies in well understood nuclear reactions which provide us with an accurate means of measuring great periods of time. Methods for determining the 'ages' of rocks use the transformation, at a known rate, of radioactive 'parent' isotopes of chemical elements into non-radioactive 'daughter' isotopes of the same, or sometimes different, elements.

In the early 1950s, US scientists led by Claire Patterson first measured isotopes of lead in meteorites and the Earth. Two of the four isotopes of lead are produced by radioactive decay of uranium. A third results from the radioactive decay of thorium. Uranium-235 decays to lead-207 with a half-life of 704 million years; and uranium-238 decays to lead-206 with a half-life of 4500 million years. Because each radioactive pair has its own rate of decay, it is possible to measure the 'uranium-lead' ages of meteorites and Earth rocks using only the relative amounts of lead isotopes present, without having to measure the amounts of uranium relative to lead.

Provided they have remained undisturbed, materials that formed 4500 million years ago will have retained all the lead-207 and lead-206 produced by radioactive decay since that time. These rocks will have high amounts of lead-207 relative to lead-206. Younger rocks, or those from which lead has been removed by heating, will have lower amounts of lead-207 relative to lead-206. So the ratios of the two lead isotopes (lead-207/lead-206) in an undisturbed rock marks the time since it formed.

One complicating factor is that the distribution of uranium and lead in rocks is not even. Uranium has a strong tendency to concentrate in oxygen-bearing minerals such as oxides and silicates, but lead favours minerals containing sulfur. If a separation of oxygen-bearing and sulfide minerals occurred 4500 million years ago, all the lead produced by radioactive decay would now be concentrated in the minerals rich in uranium.

To help resolve differences between materials, scientists measure the amounts of lead produced by radioactivity relative to a stable isotope of the same element. For the uranium-lead system, 'radiogenic' (derived from radioactive decay) isotopes of lead are compared with the stable isotope lead-204.

Measuring the ratios of the two radiogenic isotopes of lead in meteorites, Patterson found that the mineral troilite (iron sulfide) from the Canyon Diablo iron meteorite contained the lowest ratios measured at that time. Later, ratios of lead isotopes from three stony meteorites were measured. When these were plotted with measurements from Canyon Diablo troilite on a graph, they all lay on a straight line — called an 'isochron' (from Greek *isos* equal, and *chronos* time). Isochrons express the age of materials. The slope of the line is proportional to the time the materials formed, and for the meteorites studied gives an age for all close to 4550 million years. When scientists measured isotopic ratios of lead in modern volcanic and sedimentary rocks on Earth they lay on, or close to, the same line, confirming that the is Earth is also about 4550 million years old.

A sniff of oxygen

If radioactive isotopes help us date materials, stable isotopes, particularly of oxygen, tell us about the sources and conditions of materials from which rocks formed. There are three stable isotopes of oxygen (O), with atomic weights of 16, 17 and 18 that occur in the approximate proportions 99.8, 0.04 and 0.2 per cent, respectively. Because of the abundance of oxygen on Earth and its importance in rock formation, analysis of these stable isotopes provides important information helping scientists to determine relationships between many natural materials.

When organisms in coral, for example, use calcium, carbon and oxygen dissolved in seawater

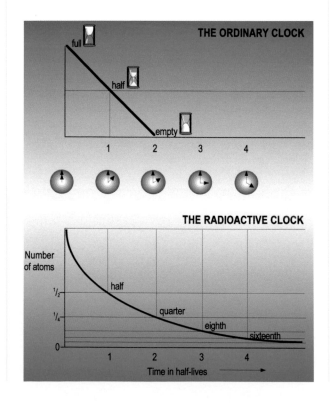

Uranium-235 and uranium-238 decay to form lead-207 and lead-206 respectively. The different half-lives at which this occurs dictates that 4500 million years ago there were twice as many uranium-238 atoms as now, while there were about 80 times the number of uranium-235. Today uranium-235 is nearly extinct, having almost completely decayed to lead-207. But uranium-238 is still producing lead-206 at about half the rate it did 4500 million years ago. The relative amounts of lead-207 and lead-206 being produced therefore change with time.

In an hourglass, the second half of the sand drains from the top in the same time as the first half. In a radioactive clock, the atoms change exponentially. If half the atoms decay in the first hour (a half-life of 1 hour), half of the remaining atoms (that is one quarter) change in the next hour, and so on.

uranium-235

92 protons
143 neutrons

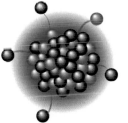

10 protons lost
18 neutrons lost

lead-207

82 protons
125 neutrons

uranium-238

92 protons
146 neutrons

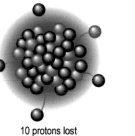

10 protons lost
22 neutrons lost

lead-206

82 protons
124 neutrons

any process like mineral growth which separates the heavier isotopes compared to light oxygen-16 has twice the effect on oxygen-18 than it does on oxygen 17. If coral contains two parts per thousand of oxygen-18 more than seawater, then it will only contain one part per thousand more of oxygen-17. In materials formed on Earth, the two heavier isotopes of oxygen always vary by the ratio of 1:2.

Conventionally, measurements are expressed as deviations (deltas or δ) from those found in standard mean ocean water. When the ratios of heavier isotopes of oxygen to light oxygen 16 in different samples from Earth are plotted against each other they lie on a line passing through standard water with a slope of exactly one half. All Earth samples lie on this line, showing that they all come from the same source material. So do all samples from the Moon. This is strong evidence that the Earth and the Moon are not chance associates, but formed from the same oxygen source in the same region of the Solar System. The importance of this for planetary scientists is that any samples formed from another source of oxygen would lie on different lines

to make their homes out of the mineral calcium carbonate, the proportions of the three oxygen isotopes in the resulting reef differ from those in the seawater. The extent of the difference depends on the temperature of the water when the coral was made. This allows geologists to measure seawater temperatures throughout the history of the building of the reef, so monitoring any climatic change. Earth scientists use the same means to obtain the temperatures at which igneous and metamorphic rocks form.

Oxygen-17 is one neutron heavier, and oxygen-18 is two neutrons heavier, than oxygen-16. Therefore

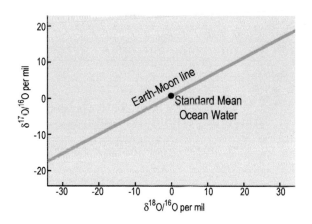

The lead-lead isochron is produced by plotting the ratio of isotopes lead-207 to lead-204 against that of lead-206 to lead-204. Points lying on the straight line have the same ratio of lead-207 to lead-206 produced by radioactive decay. The slope of the line gives an age of 4560 million years, whereas the position along the

line is determined by the ratio of uranium to lead in the sample. The igneous meteorite Sioux County has a high uranium and a low, mostly radiogenic, lead content. Earth's lead plots close to the point for troilite from the Canyon Diablo iron meteorite, showing that the planet is as old as most meteorites.

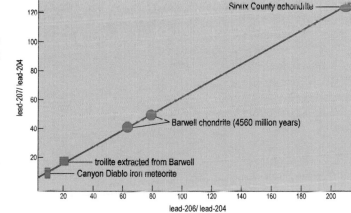

The ratios of isotopes oxygen-17 to oxygen-16 and of oxygen-18 to oxygen-16 from samples from the Earth and Moon expressed as deviations (δ) in parts per thousand (/mil) from standard mean ocean water (SMOW) define a line with a slope of one half. Samples formed from a different isotopic mixture would lie on different lines.

The planets

Leaving the Earth and taking a wider view, following the success of the manned Apollo landings on the Moon, over the last 30 years a barrage of unmanned space missions have probed the inner and outer Solar System. Some, like the Viking, Pathfinder and Venera probes, landed on Mars and Venus. Others like Pioneer and Mariner made close approaches to planets recording as many remote measurements as their equipment allowed.

Despite the wealth of information gathered from the planets, estimating their overall compositions is even more difficult than for the Earth, and there is a great deal of uncertainty. Although the inner planets are all rocky bodies, they differ in mass and density. For the three innermost planets including Earth, a general increase in mass with distance from the Sun is matched by the length of time they appear to have had active volcanoes. While Earth is still volcanically active, Venus with a similar size and density appears to have been very active until recently but now appears to be dead.

Closest to the Sun, Mercury is the smallest of the inner planets and its density is among the highest of any planet. Three quarters of its bulk is taken up with an iron-rich core, the detailed composition of which is unknown. Mercury has a magnetic field which is similar to, but less than that of Earth, suggesting that it may have a liquid outer core. Firmly in the grip of the Sun, Mercury is slung in an orbit inclined by 7° from the plane of the Earth's orbit. Mercury has a thin crust of silicates — much thinner than its large core should support. The loss of a large part of its crust, possibly by impact shortly after it formed, has been suggested as an explanation for this anomaly.

The low density of craters on the surface of Venus indicates that it has a young crust, consistent with recent volcanic activity. Basalt lavas on its surface are similar to those erupted on Earth, suggesting that it has a similar mantle. Although the planet has a similar size and density as the Earth, there are significant differences between Venus and the Earth. Venus spins very slowly in an opposite sense to the Earth with one revolution taking 243 Earth days. Its atmosphere is also very different, being made mostly of carbon dioxide. As to Venus' composition, measurements suggest that it has a core perhaps only slightly smaller than the Earth's, and that the overall make-up of the two planets is also very similar.

Mars is smaller than the Earth. This is the legacy of a deprived infancy: Mars' stunted growth results from the robbery of building materials by its giant neighbour, Jupiter. Mars' mean density, much lower than the Earth's, suggests that it also has a very different composition. Although its volcanoes now appear to be extinct, or possibly dormant, Mars has the largest known volcano in the Solar System. The Martian crust is made of iron-rich basaltic rocks and andesites. No details of Mars' interior are known with certainty. Indications from physical measurements, such as the planet's moment of inertia, are that Mars is differentiated with a dense core.

The giant outer planets, Jupiter, Saturn, Uranus and Neptune are made of gas, rock and ices. Jupiter is richer in gas compared with Saturn, Uranus and Neptune that are generally icy planets. The amount of gas relative to ice plus rock decreases from Jupiter to Neptune. Jupiter has an inner core of rock and ice surrounded by an outer core of metallic hydrogen, which is itself mantled by gaseous hydrogen and helium. Uranus, a true ice giant, may have a rocky core surrounded by an icy shell of ammonia, methane and water ices, with a covering of hydrogen and helium gas and ice. If Pluto is ignored as a rogue rock and ice planetesimal, Neptune marks the outer limit of the Solar System. Neptune contains more rock and ice than Uranus, but less gas.

Explanations for the existence of the giant planets suggest that the formation of abundant ices and their rapid growth into large planetary cores, equivalent to more than ten times the mass of Earth, allowed Jupiter and Saturn to capture much hydrogen and helium gas during the early stages of the formation of the Solar System. Uranus and Neptune failed to grow quickly enough to capture much gas, remaining essentially ice giants.

This quick tour of the Solar System illustrates one of its most important features — all the planets are different. So were the original materials from which they are made. Because meteorites have been largely unaltered since they formed, the stuff that most of them are made of represents such planetary ingredients.

The surface of Mercury viewed from Mariner 10 spacecraft in 1974. Well-preserved and densely packed craters suggest that there has been little recent volcanic activity on Mercury.

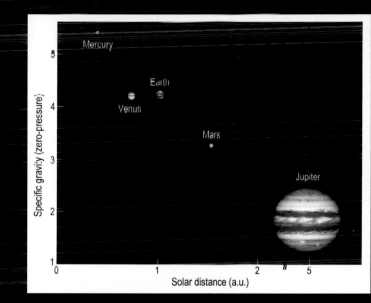

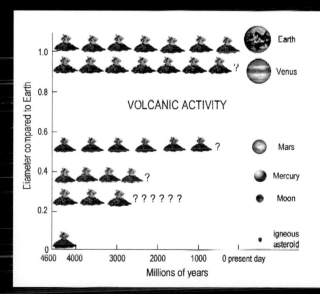

Above: The pressure relieved densities and distances from the Sun (in astronomical units) for the inner planets and Jupiter. Mercury is only one eighteenth the mass of Earth, while Jupiter is 318 Earth masses. Venus and Earth have very similar masses and densities.

Below: The persistence of igneous activity on the inner planets, the Moon and the parent asteroids of some meteorites, as measured by the density of craters on volcanic surfaces and the radiometric ages of samples, corresponds to the size of the body: the bigger, the longer.

Olympus Mons, the immense volcano on Mars, is clear testimony to recent volcanic activity on the planet. The cone is 22 km high and about 500 km across.

Jupiter, with its 16 moons, is the gas and ice giant of the Solar System.

I therefore argue that some at least of the

constituent particles of meteorites were

originally detached glassy globules,

like drops of a fiery rain.

HENRY CLIFTON SORBY, 1877

BUILDING BLOCKS OF PLANETS?

Just like any other rocks, meteorites are mixtures of minerals, but these are generally in very different amounts and associations to those seen in rocks at the Earth's surface. Meteorites are dominated by two main types of minerals: silicates rich in either the elements silicon, oxygen, iron and magnesium (like olivine and pyroxene) or in silicon, oxygen, calcium and aluminium (like feldspar); and metallic minerals made of iron with variable amounts of nickel. Around 295 different minerals are known in meteorites, including a few that are extremely rare or else unknown on Earth. Only about eight minerals in varying proportions, however, make up 90 per cent of meteorites. Importantly, no chemical element has yet been found in meteorites that is not known on Earth.

Meteorite diversity

Historically, meteorites have been divided into three categories depending on the amounts of silicate and metallic minerals they contain. 'Iron' meteorites are composed principally of metal; 'stony' meteorites (often called 'stones') consist mainly of silicates but also contain some metal; and 'stony-irons' contain metal and silicates in roughly equal amounts. By eye, or with a simple hand-lens, it is possible to slot most meteorites into one or other of these rudimentary groups.

All known meteorites are irons, stones or stony-irons, but this simple description actually conceals the diversity of materials falling to Earth, implying close family relationships where none appear to exist. Two outstanding examples are the Bencubbin meteorite found in Western Australia in 1930, and the Lodran meteorite that fell in Pakistan in 1868. Outwardly these rocks are undisputedly 'stony-irons' but as we shall see later their compositions and textures reveal ancestries closer to the stones.

Just as the peoples of the world can look very different, the genetic make-up of our species, *Homo sapiens*, shows that we are related to common ancestors. One of the aims of the modern study of meteorites is to group together 'genetically' related meteorites, seeking relationships between families that may reveal their origin. Over the last 20 years, using this approach, there have been great advances in the understanding of meteorites. Within each of the

A cut and polished slice of the Springwater stony-iron meteorite (pallasite) which was found in Canada in 1931 reveals a sponge-like network of iron-nickel metal (bright) enclosing crystals of olivine (dark green).

three broad categories, numerous distinct groups of meteorites are recognised, differing in chemistry, mineral make-up and texture.

The largest group of stony meteorites are the chondrites (pronounced *kon-drites*). Eight out of ten meteorites seen to fall are chondrites. Chondrites contain closely packed, near-spherical beads, called chondrules (from Greek *chondros* meaning grain), from which the group name originates. Chondrules are made mainly of silicates, and vary from sand-sized particles up to objects the size of a peanut. All together, chondrules can make up more than three-quarters of the meteorites in which they occur. Chondrules are unknown in any rocks from Earth, and scientists have argued passionately for more than a century over exactly how, and from what, these enigmatic objects formed.

Chondrites are extremely old rocks. Many have remained essentially unaltered since their formation around 4555 million years ago. As samples from minor planets, they record a wide variety of events that took place very early in the history of the Solar System. Considering he made the comments 123 years ago, with only a microscope as an aid, Henry Sorby's 'drops of a fiery rain' was an outstanding explanation of chondrules. Today there is general agreement that chondrules formed by rapid cooling of once molten, or partly molten, droplets of rock. What caused the melting, where the melting occurred and the nature of the starting materials are still subjects of intense debate. Importantly, chondrules were among the earliest solid materials to form in the Solar System. Because of their abundance in meteorites, understanding the origin of chondrules and their assembly to make the chondrites is fundamental to our understanding of how materials (and ultimately planets) formed in the infant Solar System.

A cross-section of the Wiluna ordinary chondrite that fell in Western Australia in 1967 contains abundant millimetre-sized beads of stony mineral (chondrules). Ordinary chondrites are the most common type of meteorite seen to fall. Different illumination (right) reveals that bright flecks of metal form a substantial portion of the mass.

NON-CHONDRITIC METEORITES

IRON METEORITES

Irons	
IAB*-IIICD*	- Mundrabilla
IC	- Mt Dooling
IIAB	- Veevers
IIC	- Kumerina
IID	- Needles
IIE*	- Miles
IIF	- Repeev Khutor
IIIAB	- Henbury
IIIE	- Paneth's Iron
IIIF	- Nelson County
IVA*	- Gibeon
IVB	- Hoba
*Irons with silicate inclusions	
Pallasites	- Imilac
	- Springwater

Mesosiderites	- Mt Padbury
	- Pinnaroo

BASALTIC ACHONDRITES

Eucrites	- Millbillillie
	- Camel Donga
	- Juvinas
Diogenites	- Johnstown
Howardites	- Kapoeta
Angrites	- Angra dos Reis

ENSTATITE ACHONDRITES

Aubrites	- Mayo Belwa
	- Cumberland Falls

PRIMITIVE ACHONDRITES

Acapulcoites	- Acapulco
Lodranites	- Lodran
Brachinites	- Brachina
	- Reid 013
Ureilites	- Novo Urei
	- North Haig

PLANETARY ACHONDRITES

Lunar achondrites, basalts, breccias & anorthosites	- Yamato 791197
SNC (Martian achondrites)	
	- Nakhla
	- Zagami
	- Shergotty
	- Chassigny

Generally, chondrites have overall chemical compositions close to the condensable portion of the Sun. This may seem a strange statement considering the bulk of the Sun is made of gaseous hydrogen and helium. By this we mean that the chemistry of some meteorites is very close to the 1 per cent of the Sun that, if it cooled, would condense into solid material.

Current theories for the origin of the Solar System invoke an initial dense cloud of gas and dust — the solar nebula — from which the Sun formed, with the material that formed the planets condensing and accreting at about the same time. The approximately 'solar' compositions of chondrites indicate that they have not undergone melting and separation after their formation. It is therefore thought that chondrites are essentially aggregates of materials retaining physical and chemical records of some of the earliest events in the history of the Solar System.

CHONDRITIC METEORITES

CARBONACEOUS CHONDRITES

CI	- Ivuna
CM	- Mighei (CM2)
	- Murray (CM2)
CO	- Ornans (CO3)
CV	- Vigarano (CV3)
	- Allende (CV3)
CK	- Karoonda (CK4)
	- Cook 003 (CK4)
CR	- Renazzo
CH	- ALH 85085

R	- Rumuruti (R3-6)
	- Carlisle Lakes (R3)

K	- Kakangari (K3)
	- Lea County 002 (K3)
	- LEW 87232

ORDINARY CHONDRITES

LL	- Yalgoo (LL5)
	- Hamlet (LL4)
L	- Barratta(L4)
	- Woolgorong (L6)
H	- Wiluna (H6)

ENSTATITE CHONDRITES

EL	- Happy Canyon
EH	- Abee

Right: The millimetre-sized chondrules extracted from the crumbly matrix of the Saratov ordinary chondrite that fell in Russia in 1918 are generally, but not exclusively, spherical.

The cut mass of the Bencubbin meteorite from Western Australia shows an internal mixture of abundant metal and silicate, suggesting it is a stony-iron. However, chemically the meteorite is much closer to the chondrites.

The Barratta ordinary chondrite found in New South Wales in 1845 shows abundant sub-circular chondrules.

A much rarer group of stony meteorites, the achondrites, include stony meteorites differing chemically and texturally from the chondrites and, as their name suggests, lack chondrules. They are usually devoid of metallic iron, and many achondrites resemble volcanic and igneous rocks on the Earth and the Moon, or their debris.

Broadly, irons, stony-irons and achondrites represent the end products of heating, melting, differentiation and mixing of more primitive Solar System materials. Although the nature of inter-family relationships between the chondrites and other meteorite groups is uncertain, the differentiated meteorites may have formed in a number of asteroids by the melting of chondrite-like materials and the gravitational separation of molten metal and silicates. Thus the irons largely represent solidified and slowly cooled separations of once molten metal, whereas achondrites are more or less recycled samples of the predominantly silicate residues. Stony-irons may have formed either by gentle or violent mixing of both solid and liquid of both metal and silicate at various depths within their parent asteroids.

Leaving their outward features aside, in reality there are only two main categories of meteorites: those that contain chondrules and those that, for a variety of reasons, do not. As knowledge of meteorites increases, the 'genetic' approach to meteorite families is making the historic grouping of iron, stony and stony-iron meteorites almost redundant

FALL AND FIND FREQUENCY

Records of meteorite falls give a measure of the relative abundances of the different types. The 'fall frequency' is the percentage of each type of meteorite observed to fall. The total number of chondrites and achondrites in collections of meteorites throughout the world (that is observed falls plus finds from unobserved falls) correspond well with those predicted by their fall frequency.

In contrast, irons and, to a lesser extent, stony-irons, which are the rarest types seen to fall, appear to be over-represented in collections. The reason for the abundance of iron and stony-iron finds is that, by virtue of their exotic nature, the strongly metallic meteorite types are most easily recognised.

Under the microscope, a thin slice through a chondrule less than 0.5 mm across from the Saratov ordinary chondrite shows that it was originally molten at around 1400°C. From the rapid cooling of the liquid, some feathery crystals of olivine grew quickly but the rest of the droplet chilled to brown glass. Textures like this are typical of rapid solidification and cooling.

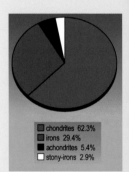

chondrites 62.3%
irons 29.4%
achondrites 5.4%
stony-irons 2.9%

The actual proportions (above) of the different meteorite types found differ from what would be expected from their frequency (below) as falls. This is because some types are more easily recognised.

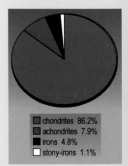

chondrites 86.2%
achondrites 7.9%
irons 4.8%
stony-irons 1.1%

Achondrites lack chondrules, are generally devoid of metal, and have textures and chemistries similar to igneous rocks. The main mass of the Millbillillie achondrite, from the shower that fell in Western Australia in 1960, shows a texture of a jumble of angular fragments welded together ('breccia').

The chondrites

A microscopic view of chondrites shows they are made of intricate, varying and complex mixtures of silicate minerals containing iron and magnesium, metallic iron-nickel and sulfur-rich minerals such as troilite (iron sulfide). Excluding those elements like hydrogen, carbon, nitrogen and sulfur that are easily lost through heating, overall the chemical make-up of meteorites containing chondrules is remarkably similar: the elements iron, magnesium, silicon and oxygen make up more than 90 per cent of their total weight. However, the total amount of iron they contain, and the form in which the iron occurs, varies between chondrites such that, on the basis of this and other chemical and isotopic differences, around 14 distinct families are recognised.

The largest group, collectively known as the 'ordinary chondrites', accounts for more than half of all known meteorites. Although 75 per cent of their bulk is made of silicate minerals, ordinary chondrites contain substantial amounts of iron both in the form of silicates containing iron, such as olivine and pyroxene, and as metal and sulfide. Three *chemically* distinct families make up the ordinary chondrites: the H or 'high-iron' chondrites, the L or 'low-iron' chondrites, and the LL or 'low total iron, low metallic iron' chondrites. The H-chondrites contain

The chondrite meteorite families. Type 7 chondrites are those in which chondrules have been erased.

the most metallic iron, but less iron combined with oxygen, whereas the L- and LL-chondrites contain less metal and a higher proportion of oxidised iron. The relationship between the three ordinary chondrite families is not simply an equitable redistribution of their iron and oxygen budgets, for the total amount of iron decreases between H- to L- to LL-chondrites.

Two other rarer groups of chondrites, 'enstatite chondrites' and 'carbonaceous chondrites', represent extremes in composition. The enstatite chondrites (designated E) are poor in oxygen, rich in iron and contain abundant metal and sulfides. The main silicate mineral they contain (enstatite) that lends its name to the group, is a magnesium silicate containing no iron. In contrast, carbonaceous chondrites (designated C) generally contain little or no metallic iron, but their silicates are mostly iron-rich.

Within each of the ordinary and other chondrite families, meteorites show varying amounts of mineral growth, or recrystallisation, that has gradually erased their chondrules. These metamorphic changes took place during prolonged gentle 'cooking' of the rocks at varying temperatures up to maximum of about 950°C — well below their melting points of around 1500°C.

Four main *textural* types of ordinary chondrite are assigned labels on a numbered scale from least crystallised (type 3) to most (type 6) with types 4 and 5 intermediate to these extremes. Type 3 chondrites show prominent chondrules made of well-formed or feathery crystals in glass, set in a fine-grained matrix, and their minerals vary widely in composition. In the well-cooked type 6 chondrites, mineral growth has all but erased their chondrules which now appear as ghostly outlines, and their minerals are generally uniform in composition.

Using chemical and textural similarities it is possible to 'pigeon-hole' most ordinary chondrites. The Binningup meteorite, for example, is a high-iron chondrite with a crystallised texture: showing that it is an H5. Moderately to severely crystallised ordinary chondrites of types 5 and 6 belonging to the H- and L- groups, are the commonest meteorites to fall.

Chemical Groups		Types
Carbonaceous (C)	CI	1
	CM	1-2
	CR	2
	CO	3
	CV	2-3
	CK	3-6
	CH	
Ordinary (O)	H	3-7
	L	3-7
	LL	3-7
Enstatite (E)	EH	3-6
	EL	3-6
R (Rumuruti)		3-6
K (Kakangari)		3

BITS
AND
PIECES

Chondrules are abundant constituents of most chondrites and there are many kinds of chondrules with different chemistries and textures. In those chondrites which escaped alteration after they were assembled, pristine chondrules record important information about the nature of the materials from which they were made, and the brief history of their formation.

As reflected in the overall composition of the chondrites, the greatest number of chondrules are rich in iron and magnesium, containing abundant crystals of olivine and/or pyroxene. For elements other than metal-loving, siderophile, and temperature sensitive, volatile elements, their overall chemical make-up is similar to the CI chondrites. Between the high temperature calcium-aluminium rock fragments and iron-magnesium rich chondrules there is a wide range of bulk chondrule compositions. Chondrule chemistry also varies between the families of chondrites. For instance, chondrules with chemistries between the high-temperature inclusions and the iron-magnesium rich varieties are generally aluminium-rich.

As with their chemistries, there is a great variety of textures among chondrules. Laboratory experiments to 'manufacture' chondrules reveal that different amounts of melting, the length of time they were molten and the rate at which they solidified, all have some bearing on their resultant textures. Solidified chondrules cooled at rates ranging from a few degrees per hour to around hundreds of degrees per hour before being bound into their host rocks.

Cut, mounted on glass slides, and illuminated under the microscope, two ordinary chondrites of the same family show markedly different textures. The H3 chondrite (left) shows prominent chondrules set in fine-grained matrix. Heating of the H5 chondrite (Binningup, below) has caused the minerals to recrystallise, and chondrules are barely discernible. These variations in texture reveal some of the processes that took place while the rocks were resident in their parent asteroids.

Common minerals occurring in meteorites

IRON–NICKEL METAL Mainly represented by two alloys containing variable amounts of iron, nickel and cobalt: kamacite, with less than 7 per cent nickel; and taenite usually with more than 25 per cent nickel. Metal occurs in most meteorites and is the main constituent of irons.

OLIVINE An iron-magnesium silicate. Major constituent of most chondrites and pallasite stony-irons. Rare in other stony-irons, achondrites and irons. Gem quality olivine is known as peridote.

PYROXENE A group of predominantly iron-magnesium-calcium silicates. Major constituent of most chondrites, achondrites and mesosiderite stony-irons.

PLAGIOCLASE FELDSPAR A silicate of sodium-calcium-aluminium. Occurs in most chondrites, achondrites and mesosiderite stony-irons.

TROILITE Iron sulfide. Present in varying amounts in most types of meteorite.

'SERPENTINE' A group of iron-magnesium silicate minerals containing water which form by the alteration of other iron-magnesium silicates. Abundant in some carbonaceous chondrites. Rare in all other meteorites.

MAGNETITE A magnetic oxide of iron. Common in some carbonaceous chondrites. Generally rare in other meteorites.

No single theory has so far explained all the features of chondrules. In carbonaceous chondrites such as Allende, chondrules vary widely in their contents of major elements compared with the bulk composition of the meteorite. This chemical variation puts some constraints on the ways in which chondrules formed. Large variations in magnesium, aluminium, calcium, iron and some trace elements in chondrules suggest that they formed from variably mixed starting materials.

Chondrites are mixtures of chondrules and, in the least altered type 3 chondrites, a generally dusty medium termed 'matrix'. The make-up of this pristine matrix can vary on the scale of thousandths of a millimetre revealing important information about its formation. The relationship between chondrules and their embedding medium is not a simple one. Generally, variations in the composition of matrix are due to differing amounts of iron-rich olivine and other matrix minerals, such as iron-poor pyroxene together with material rich in silicon, aluminium, calcium, sodium and potassium.

Matrix and chondrules are essentially complementary in composition, and it is not easy to make chondrules simply from matrix-like materials. Relative to chondrules, matrix is rich in iron oxide. Chondrules typically contain less than a fifth of the whole meteorite abundances of metal- and sulfur-loving elements, while they are not impoverished in volatile elements relative to the bulk chondrite. These variations in chemistry suggest that chondrules formed from material that had already undergone depletion of volatile elements and some separation of metal, sulfide and silicates. Although there is considerable variation in chondrule compositions even within groups, the compositional differences between chondrite groups are due to variations in the iron-rich matrix and metal. Importantly, the differences in chondrite composition show that after chondrule formation there was incomplete mixing of material.

Not all minerals in chondrites formed at high temperatures. *Carbonaceous* chondrites are often dark rocks, rich in complex carbon compounds, and some also contain water-bearing minerals, such as clays and the minerals epsomite (epsom salts: hydrated magnesium sulfate) and gypsum (hydrated calcium sulfate). These formed at low temperatures by the alteration of other minerals.

Most carbonaceous chondrites were untouched by the heating that caused recrystallisation of the ordinary chondrites, and they account for types 2 and 1. Instead of heating, these stages represent progressive alteration of their minerals by fluids. Some carbonaceous chondrites have been slightly heated, and a few are severely recrystallised, but the great majority retain a nearly undisturbed record of their history. Even the activity of water that altered their minerals did not change their overall chemistry.

Carbonaceous chondrites are often rich in carbon which gives them their name. Some, however, are carbon-poor and the abundance of this element is not the main characteristic of carbonaceous chondrites. The common traits of these meteorites are their higher abundances of magnesium, calcium and aluminium relative to silicon, than those of ordinary chondrites. Carbonaceous chondrites are also rich in oxygen, consequently their silicate minerals are mostly iron-rich, and they generally lack metal. There are seven families of carbonaceous chondrites: designated CI, CM, CO, CV, CK, CR and CH.

The CI1 chondrites are extremely rare with only a few examples known. They are typified by the Ivuna (thus I) meteorite that fell in Tanzania in 1938. The largest known example is the Orgueil meteorite that fell in France in 1864. These meteorites are made largely of minerals containing appreciable amounts of water. Indeed, mineral-bound water alone can account for up to a fifth of their total weight. Magnetite (magnetic iron oxide) is the second-most abundant mineral in CI chondrites but they also contain carbonates, sulfates and sulfides. Paradoxically, these meteorites lack chondrules, but overall chemical similarities with meteorites containing chondrules and the presence of rare grains of high-temperature minerals such as olivine and pyroxene which may have come from chondrules show that they are 'chondrites'.

Significantly, with the exception of gases such as hydrogen, helium, oxygen and nitrogen, CI chondrites have chemistries that match the Sun very closely. Since the Sun accounts for 99.9 per cent of the mass of the Solar System, these CI chondrites are

the best samples of 'average' Solar System material that can be studied in the laboratory. Most theories for the origin and evolution of the solid materials of the Solar System start from the assumption that the composition of the whole was originally similar to this 'primitive' family of meteorites.

The most abundant carbonaceous chondrites are the CM2 chondrites, named after the Mighei meteorite that fell in the Ukraine in 1889. These meteorites contain some chondrules, but they make up only about 12 per cent of their volume. The chondrules sit in a medium that is very similar to the bulk of CI chondrites, made largely of water-bearing minerals with some magnetite. Carbon is abundant as complex carbon compounds and also carbonates, such as calcite (calcium carbonate). Half the bulk of CM chondrites is made of minerals that formed at high temperatures: those in the chondrules, tiny isolated crystals of olivine and pyroxenes, and loose clumps of these minerals together with a calcium- and aluminium-rich glass, metal and other minerals such as spinel (magnesium aluminium oxide). The intricacies of the make-up of CM chondrites tell a complex story of assembly and partial alteration of their constituent minerals by fluids.

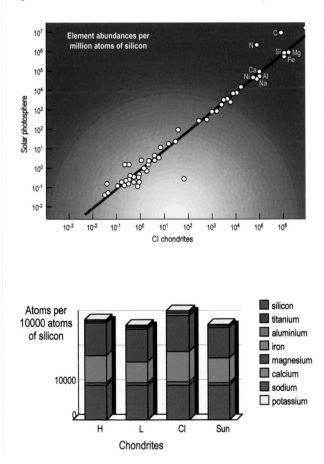

Abundances of the elements in CI chondrites compared to those in the Sun's photosphere (per 1 million atoms of silicon). Apart from a few volatile elements, there are similar proportions for most elements.

The difference in composition between H and L ordinary chondrites, CI chondrites, and the Sun, is partly a different proportion of iron compared to silicon. The process that enriched, or depleted, iron relative to silicon is called metal/silicate fractionation.

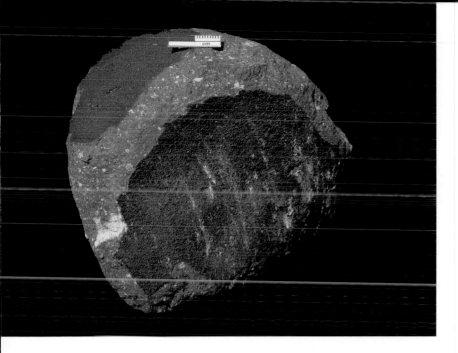

The remaining carbonaceous chondrites (CO, CV, CK, CR and CH) are all generally low in carbon. The CO chondrites all belong to type 3, but their minerals show evidence of small variations in exposure to heat after their assembly. The family is named after the Ornans meteorite that fell in France in 1868, about 5 kg of which is preserved in museums. CO chondrites are packed with small chondrules, usually less than 0.2 mm across, which together with grains that may be chondrule fragments make up nearly half of their volume. Like the CM chondrites, COs contain some water-bearing minerals, but their content of carbon compounds is less than 1 per cent. Other minerals in CO chondrites are metal, magnetite and sulfides.

The 16-kg carbonaceous chondrite that fell at Vigarano in Italy in 1910 gives its name to the CV chondrites. The great majority of CV chondrites (more than 30) are type 3: only one stone from the Nullarbor, Mundrabilla 012, is a CV2. These meteorites have large chondrules and abundant fine matrix between chondrules. On cut surfaces, their most striking features are irregularly shaped, white rock fragments often the size of grapes. The bright appearance of these fragments demands attention, and indeed they have been the most intensely stud-

ied parts of meteorites, for the minerals they contain are some of the oldest in the Solar System and hold part of the key to an understanding of its birth.

Encapsulated in these rock fragments are minerals rich in the elements calcium, aluminium and titanium, and also metallic grains rich in platinum and iridium. Minerals that contain these elements form at very high temperatures, above 1500°C. One theory is that they condensed directly from the gas and dust cloud that ultimately formed the Sun and planets. These important components of meteorites are not exclusive to the CV chondrites as they occur in many other groups, notably the CO chondrites, but they are nowhere larger than in CVs.

In November 1930 more than 40 kg of a hitherto unknown type of meteorite fell at Karoonda in South Australia. Thought for a long time to be a CO chondrite, recent finds of meteorites in the Nullarbor and Antarctica similar to Karoonda have established a new group: the CK chondrites. These meteorites are highly oxidised, contain no metal, and their silicate minerals are iron-rich. Magnetite and sulfides account for some of the rest of their make-up. They are the only family of carbonaceous chondrites with members of types 3 to 6, with most having been partly recrystallised. The chemistry of CK chondrites is similar to CO and CV chondrites but significant differences set them apart.

The CR and CH chondrites differ from the other groups in that they both contain substantial amounts of metal. The CH chondrites are not widely accepted by scientists as a true family group. Breaking with tradition, the H stands for 'high-iron' as up until recently there was no observed fall of their kind. An observed fall in Nigeria in 1984, named Gujba, may have provided the first fresh example of the group. An Antarctic CH find, ALH 85085, is one of the most metal-rich and sulfide-

Far left: Abundant metal and an extreme poverty of oxygen are characteristic of enstatite meteorites, such as Abee, which fell in Canada in 1953.

Left: The small Carlisle Lakes meteorite found on the Nullarbor Plain, like other Rumuruti chondrites, appears to be a sample from a region of the early Solar System different to all other chondritic meteorites.

poor chondrites known, with metal forming a fifth of the meteorite. CH chondrites have tiny chondrules (0.02 mm across) that make up around 10 per cent of their bulk, while chondrules and chondrule fragments can make up 70 per cent. Only 11 of these oddball rocks have been found to date.

CH chondrites may be related to CR chondrites, another group that has only recently been recognised. The Renazzo meteorite fell in Italy in 1824. Around 7 per cent of its volume is taken up by metal, usually associated with chondrules. Chondrules in CR chondrites are large, averaging around 0.7 mm across, and together with chondrule fragments make up more than half of the rock. All CR chondrites, including several new North African and Nullarbor finds, are type 2, showing some alteration of their minerals by water.

Two very similar and extremely unusual meteorites, Bencubbin and Weatherford, are made of more than half metal and the rest nuggets of iron-poor silicates and fragments of a number of different chondrites, some of which are carbonaceous. Descriptively they are 'stony-irons', while fragments of the chondrites enclosed within them are evidence of mixing by collisions of primitive materials early in their history. Unusual silicate textures and compositions similar to chondrules suggest that they are chondrites. Recently, Bencubbin and Weatherford have been linked to a Saharan meteorite (Hammadah al Hamra 237) and two Antarctic meteorites (Queen Alexandra Range 94411 and 94627) as a proto-family, designated CB chondrites. While their true affinities are still to be determined, their mineralogy and chemistry indicate they are closely related to CH chondrites and more broadly related to the CR clan. The CB chondrites are the most metal-rich chondrites known.

A small number of carbonaceous chondrite 'orphans' have yet to find families. Some may be distant cousins and related to the established families, while others are still waifs and strays. Their birth certificates, however, are written chemically, and these reflect their formation at different locations in the Solar System, but not at different times.

If the carbonaceous chondrites are strange rocks, then the *enstatite* chondrites are perplexing. These bizarre rocks are among the poorest in oxygen that the Earth has sampled from the entire Solar System. They have about the same budget of iron as H-chondrites, but virtually none of it is combined with oxygen. Their main silicate mineral is the iron-free form of pyroxene, enstatite (magnesium silicate) that lends its name to this family.

The oxygen starvation of enstatite chondrite formation resulted in exotic minor minerals, some of which are unknown on Earth. Calcium, magnesium and potassium are elements that would, in the presence of oxygen, normally form silicates. In enstatite chondrites, calcium and manganese have found a strange bedfellow in sulfur, forming oldhamite (calcium sulfide) and niningerite (magnesium sulfide), while potassium has linked with copper, iron, nickel and sulfur to make djerfisherite.

There are two separate clans of enstatite chondrites, a 'high-iron' group (designated EH) with 30 per cent iron by weight, and a 'low-iron' group (designated EL) with about 25 per cent. Like the ordinary chondrites, enstatite chondrites are rich in chondrules, and these are generally larger in EL chondrites. Although most have been recrystallised, both the EH and EL clans have members of types 3 to 6.

Sustained collecting in Antarctica and hot deserts has turned up a few other family groups. A small chondrite, weighing less than 50 g, found near

Carlisle Lakes on the northern fringe of the Nullarbor in 1977 was the first of its kind recognised. Gradually, other similar examples were found in Antarctica, and for many years they were called the 'Carlisle Lakes-like' chondrites. However, lying un-noticed since 1938 in the collection of the great museum of the Humboldt University in Berlin was a fragment of a similar meteorite that had fallen in 1934 at Rumuruti in Kenya. Now known as the R chondrites, the family boasts around 17 members including Carlisle Lakes.

Rumuruti chondrites are chemically different from ordinary, carbonaceous and enstatite chondrites. They are highly oxidised rocks with silicates dominated by olivine. Metal is extremely rare while nickel-bearing sulfides are abundant. With the exception of Carlisle Lakes, most R-chondrites are mixtures of related chondrite rock fragments that have been variably recrystallised.

Last in the inventory of chondrites is a proto-family named after the Kakangari meteorite that fell in 1890 in India. Together with Lea County 002 found in the USA and an Antarctic meteorite (LEW 87232), the K-chondrites differ in most respects from all other chondrites. They are made of magnesium-rich silicates and contain metal and sulfides.

Ashes to ashes, dust to dust

Since the chance discovery of magnetic spherules in red-coloured clay dredged from the ocean bottom by HMS *Challenger* (an oceanographic exploration vessel from Britain) in the 1880s, the study of 'cosmic dust' has grown in importance. Particularly important are dust particles spared the rigours of atmospheric passage. Interest increased dramatically in the early 1980s after extraterrestrial dust was trapped by high-flying aircraft.

Dust accounts for the largest amount of material reaching Earth today from space. Droplets from the melting of small particles during fiery passage through the atmosphere and the infall of interplanetary dust particles make up the recovered sample that, by comparison with the great mass of meteorites, is still unfortunately small. Most of these tiny particles are thought to come from asteroids, although a large contribution may be from comets. Because of the way dust is continually delivered to the inner Solar System and its gentle arrival on Earth, interplanetary dust particles probably sample a wider region of the Solar System than the large meteorites stored in the world's museums.

Much of the debris in the upper atmosphere owes its origin to the Earth and human activity; volcanic dust, soot from fires, dust and other flotsam generated by the 'space age'. Some particles come from burnt rocket fuels and satellite debris re-entering the Earth's atmosphere. So how is genuine extraterrestrial dust recognised? Many spherules found in sediments and ice on Earth contain olivine, magnetite and grains of metallic iron-nickel, minerals commonly found in meteorites. Most of these come from melting of meteoroids in the atmosphere, but some are unmelted and probably fell as micrometeorites.

Chemical analysis of hundreds of dust particles collected in the upper atmosphere shows that the majority have compositions similar to CM carbonaceous chondrites. Only a small number resemble the ordinary chondrites. Remarkably, there are only two kinds of interplanetary dust particles: those that have a 'chondritic' chemistry and those that do not. 'Chondritic' particles do not contain 'chondrules' for their sizes are much smaller than any known chondrule, rather the elements they contain have relative proportions similar to their cosmic abundances defined by the 'solar' composition of CI carbonaceous chondrites. Chondritic particles have been found stuck to non-chondritic particles indicating a common origin.

Chondritic particles are mostly black, sooty aggregates rich in carbon. A single particle only a hundredth of a millimetre across can be made of thousands of grains. Despite some similarities to carbonaceous chondrites, chondritic particles have textures, and sometimes mineral associations unlike any chondrite. A few resemble the fine matrix of CI

Weathering of the Cook 003 CK4 carbonaceous chondrite, found on the Nullarbor Plain, has revealed its abundant chondrules that stand out on its knobbly surface.

A comparison of the composition of CI chondrites to estimates of the Earth based on the composition of the mantle and gross structure of the Earth (I), and its uranium content and the ratios of elements to uranium in chondrites II: elements shown as atoms per 10 000 atoms of silicon). Both estimates show the Earth to be relatively richer in iron than chondrites, and poorer in sodium and potassium.

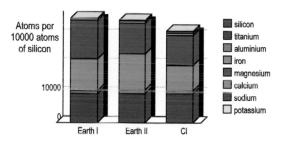

Atoms per 10000 atoms of silicon

10000

- silicon
- titanium
- aluminium
- iron
- magnesium
- calcium
- sodium
- potassium

Earth I   Earth II   CI

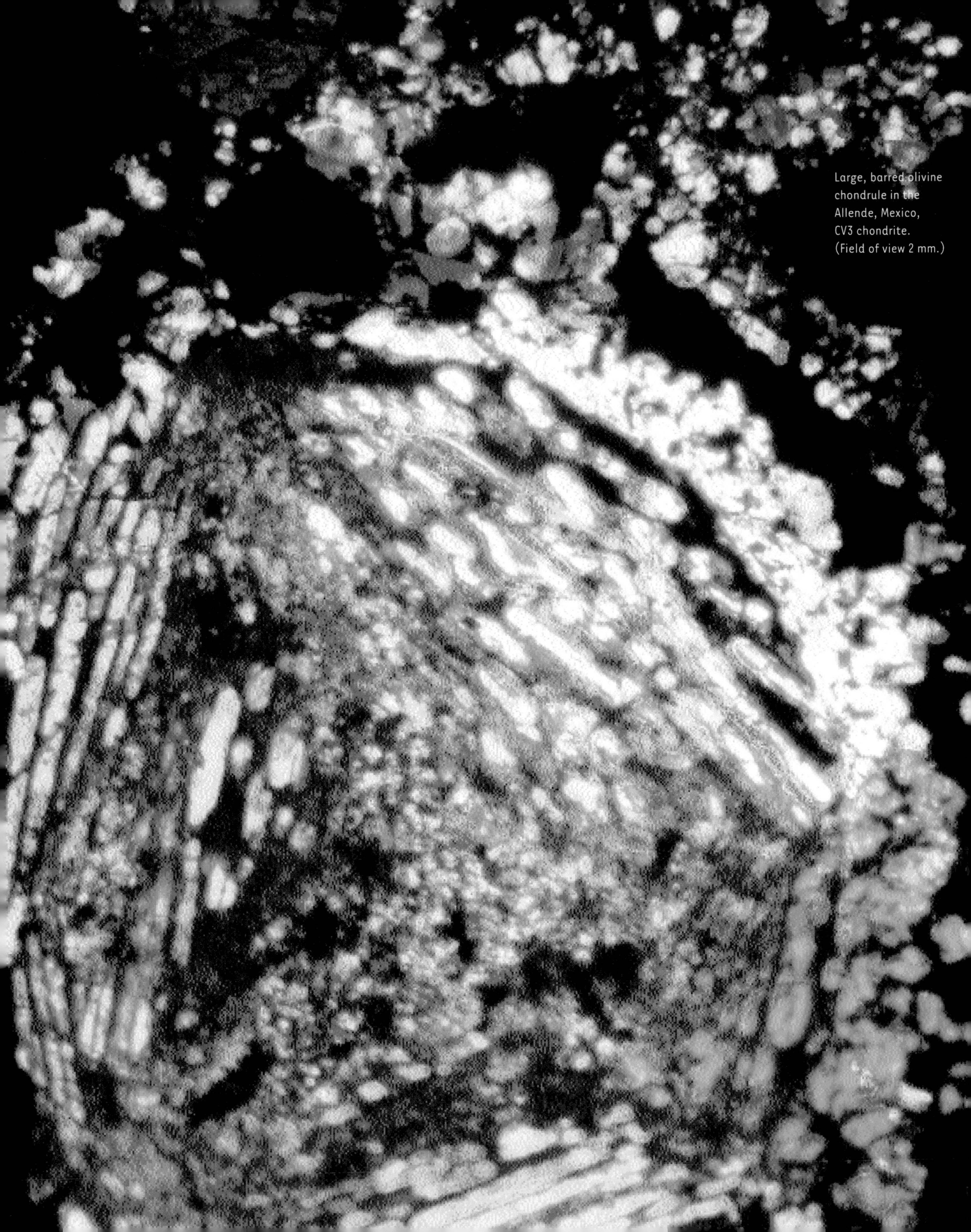

Large, barred olivine
chondrule in the
Allende, Mexico,
CV3 chondrite.
(Field of view 2 mm.)

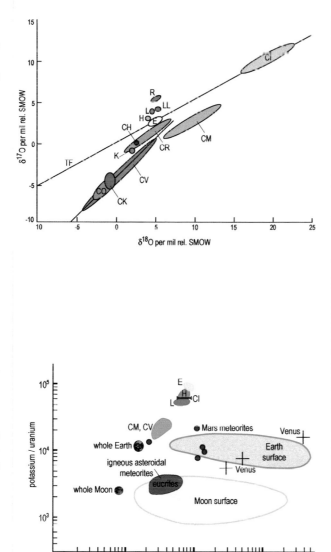

Top: Certain ratios of the different oxygen isotopes are characteristic to many chondrite families, and most groups plot in unique positions on the oxygen isotopic diagram relative to Standard Mean Ocean Water (SMOW). Carbonaceous chondrites generally plot in a region below the Earth's terrestrial fractionation (TF) line, while ordinary, enstatite and Rumuruti chondrites have well defined clusters on or above the TF line. The dry minerals of the CV group plot along a steeper line (with slope 1), which is interpreted as showing a mixing of solids rich in oxygen-16 with gases poor in oxygen-16. Variations within the CM, CI and CR groups, with shallower slopes, have resulted from isotopic separation due to varying alteration of these materials by water, probably within their parent bodies.

Below left: The abundance of potassium and uranium in the inner planets and some meteorites. The abundance of these elements in CI chondrites is taken to represent a primordial composition. The depletion of temperature-sensitive (volatile) elements, like potassium, relative to unaffected (refractory) elements, like uranium, suggests that the volatile elements may have been removed from the inner planets at an early stage.

Below: Chondrites hold a record of a wide variety of oxidation/reduction states. The relationship between iron (Fe) as oxide and silicate, and iron as metal and sulfide separates most of the various chondrite groups. Highly reduced, metal-rich enstatite chondrites (EL, EH) lie at one extreme, with highly oxidised carbonaceous chondrites (CM, CI) at the other. Meteorites of the same bulk iron content, but varying oxidation states, fall along the diagonal lines. Within the closely related ordinary chondrite groups (H, L, LL) the difference is in either the addition or subtraction of either iron or oxygen. The carbonaceous series of chondrites (CR, CO, CV, CK, CM, CI) represents increasing states of oxidation. Not plotted here are the metal-rich CH and CB chondrites.

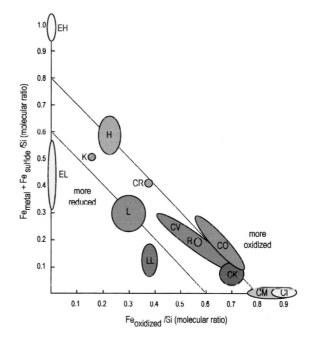

chondrites with water-bearing minerals. These interplanetary dust particles, however, have not been extensively altered by water. Non-chondritic particles are mostly olivines and pyroxenes rich in magnesium, and sulfides containing iron and nickel. Some particles are spheres resulting from the melting of fluffy aggregates in the atmosphere.

While meteorites have been altered on their parent asteroids, and subsequently damaged by their break up and delivery to Earth, dust particles have been altered by different agencies. Many particles have escaped melting, especially those that were rescued by aeroplanes before passing through the atmosphere, but all particles have suffered some heating. Other forces affecting particles include 'space

weathering'. For instance, in space, unprotected particles are bombarded by high-energy rays from the Sun. Intense radiation destroys the crystal structures of minerals, and the damage is compounded by heating during passage through the atmosphere.

Chemical fingerprints

Each of the chondrite families is unique. Their distinguishing features — like variations in the size and abundance of chondrules or calcium-aluminium inclusions they contain, their overall chemical compositions, their oxygen isotopes and the distribution of iron and oxygen between different minerals — all define their family status. Mixtures of high and low temperature components, each family carries a record of the conditions in the different parts of the Solar System where they formed.

The overall picture is not so clear. Relationships between families reflecting particular regions of the Solar System, or shared parent asteroids are generally not so evident. There is a sneaking suspicion that some families formed at different distances from the Sun, where different temperatures influenced the amount of oxygen that was available for their construction. In this way, enstatite chondrites might have formed in oxygen-poor regions close to the Sun, whilst oxygen-rich carbonaceous chondrites formed in the cooler outer reaches of the Solar System. Other features of chondrites fail to mark such family demographics.

No grand, all-encompassing picture of the formation of chondrites emerges. Differences in the outward make-up of chondrites are considerable. Abundant chondrules in ordinary and enstatite chondrites dwindle to virtually nothing in CI chondrites. What this means in the genesis of chondrites is unknown. Perhaps the manufacture of chondrules, or their collection into rocks varied from one place to another throughout the Solar System, yet the driving force responsible for this is elusive.

Differences between the chondrite families not only record events during their birth, but also the processes that took place while they were still part of their parent asteroids. Differences in degrees of heating or alteration by water mirror local conditions. The generally small variation of characters within each family of chondrites suggests they came from individual asteroids.

The Sun-like chemistry of the CI chondrites forms a base-line for Solar System materials. The CI chondrites are the richest in hydrogen, nitrogen and carbon. These elements are present in abundant carbonaceous compounds and water-bearing minerals that formed at low temperatures. When compared with the CI chondrites, other carbonaceous chondrites are generally rich in minerals that formed at high temperatures, but are relatively poor in volatile, temperature-sensitive elements. Ordinary chondrites, and Rumuruti and Kakangari chondrite families are all rich (relative to CI chondrites) in oxygen-loving elements such as magnesium and sodium, and also metal-loving elements such as iridium and nickel.

Among the most distinctive of characteristics are the meteorite families' isotopes of oxygen. As mentioned in Chapter 5, the proportions of the three stable isotopes of oxygen can tell us much about the source of oxygen from which the rocks were made. Important

in separating one meteorite family from another, oxygen isotopes also cast light on the relationship between meteorites and major bodies in the Solar System, including any alteration they may have suffered.

Most chondritic families have unique mixes of the three isotopes of oxygen (oxygen-16, -17 and -18). Diagrammatically, carbonaceous chondrites generally fall below the line defining the region of the Solar System occupied by the Earth and the Moon, whereas enstatite, ordinary and Rumuruti chondrites either lie on the Earth line, or above it. The CV, CO and CK chondrites lie on a line with a slope that is twice as steep as the line through ocean water, and the rocks of the Earth and the Moon.

This variation of oxygen isotopes in meteorites cannot be explained by a simple separation of isotopes by their masses, suggesting it is the result of mixing of oxygen from two different sources. One explanation is that carbonaceous chondrites contain Solar System oxygen spiked with almost pure oxygen-16. This overdose of the lighter oxygen isotope could not have originated in the Solar System, but might be the remnant of a supernova. All the waterless components of carbonaceous chondrites fall along the line with the slope of 1. Carbonaceous chondrites wholly or partly altered by water fall along lines with shallower slopes. These trends are attributed to the separation of the isotopes of oxygen according to their masses during alteration of carbonaceous chondrites by water on their parent asteroids.

Chondrites as a whole tell us a great deal about the infant Solar cloud. Primitive or pristine chondrite chemistries, with their varied isotopic mixes, paint a picture of the swirling cloud of gas and dust from which they formed. By comparison, their metamorphic heating, alteration by fluids, and violent mixing of chondrite types all record a turbulent episode in the evolution of their parent objects in the juvenile Solar System. What happens to chondrite-like materials when they melt is the subject of the next two chapters.

Postscript from the planets

Just as the planets are different, so are the chondritic meteorites. If these are the original planetary building materials, is it possible to construct the Earth, or any of the other inner planets, from known chondrites?

Within the small bounds of knowledge about the overall chemistry of the Earth and its near neighbours, none of the chondrite families is a match for the Earth. For instance, compared with CI chondrites, the Earth contains more iron and magnesium, a similar quantity of calcium, but very much less sodium and potassium relative to silicon.

One of the most significant features of the inner planets is their apparent loss of moderately temperature-sensitive elements, such as sodium and potassium. Out of the few chemical measurements available for all of the inner planets and chondrites is their uranium and potassium contents. Uranium is a 'refractory' element, tightly bound in the rocks in which it occurs and not easily dislodged by high temperatures.

Generally, the inner planets are poor in potassium compared to chondrites. Venus is only slightly smaller than the Earth and measurements of its radioactivity by Russian spacecraft suggest that it has similar contents of elements such as potassium and uranium. The relative amounts of potassium to uranium measured in Martian meteorites indicate a higher content of volatile elements for that planet, but still

lower than the Sun-like CI chondrites. Essentially, all the common chondrite families have unsuitably high ratios of potassium to uranium to have been the construction materials for the inner planets.

At even the most basic level of density measurements, there is no comparison between chondrites — such as the ordinary, enstatite and CV chondrites — and Mercury, Venus and the Earth. However, the density of Mars (once the highly compressed nature of deep regions of the planet is accounted for) is similar to the chondrites.

All of these differences help to limit the possibilities for the nature of the material from which the inner planets were constructed. Except for the CB chondrites, it seems that their original building blocks are not represented among the infall of meteorites to Earth. Perhaps the Earth-like meteorites were swept up by the planet as it grew in the early Solar System and are no longer to be found beyond the Earth's distance from the Sun.

Rocks,

like everything else,

are subject to change and so also are our views on them.

BY LOEWINSON-LESSING, 1936

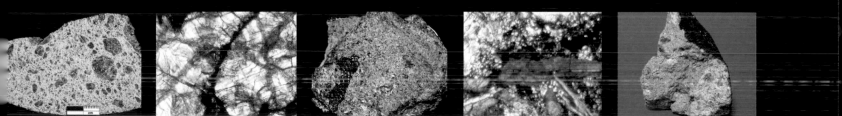

WHEN ASTEROIDS MELT

Severe heating on some parent asteroids produced a broad range of meteoritic types from chondritic precursors. Those with low degrees of melting include the acapulcoite, lodranite and brachinite achondrites. Imperfect separation of metal and silicate produced silicate-bearing irons, and extensive melting and separation gave rise to the eucrite achondrites, pallasites and irons.

Achondrites and related meteorites

Most achondrites bare clear testimony to episodes of melting on their parent asteroids in the early Solar System. Some are igneous rocks, or their debris, while others chemically resemble chondrites. Most achondrites are chemically similar to the basaltic rocks commonly erupted from volcanoes on Earth. Their textures show that most of them cooled slowly enough for visible crystals to grow.

Achondrites account for only about 5 per cent of all meteorites, and 8 per cent of meteorites seen to fall. Three kinds of achondrite — the eucrites, diogenites and howardites — are closely related and may be debris from a single asteroid. Together, these rocks represent the greatest haul of igneous material from near the surfaces of planet-like bodies beyond the Earth and Moon available for study.

Viewed under the microscope, the basalt-like *eucrites* are made of lath-like grains of plagioclase feldspar (calcium-aluminium silicate) and grains of pyroxene that grew from molten magma. Their sometimes jumbled internal appearance shows that, after they solidified, many eucrites were broken

and fractured by violent impacts on their parent asteroid. A few have undisturbed textures, typical of cumulates that formed by the slow settling and accumulation of crystals from molten magma.

Diogenites are rare achondrites made mainly of a pyroxene (magnesium-iron silicate). Like the eucrites, they also crystallised from molten magma and most have been broken and mixed by impacts. Howardites (named in honour of the pioneering meteorite chemist Charles Howard) are cemented 'soils' from the surface of an asteroid. They contain fragments of other achondrites, such as eucrite and diogenite, and also rare fragments of chondrites that are almost always carbonaceous. Their textures show that they formed by the accumulation of broken fragments produced by impacts, and they contain pieces of both the target rocks and projectiles.

Howardites have also been intensely bombarded by the Solar Wind. These charged particles are

Right: Microscopic examination of the Millbillillie eucrite, which fell in Western Australia in 1960, shows laths of plagioclase feldspar (white) intergrown with pyroxene (dark). This texture is typical of rocks that form from molten magma. (Field of view 4 mm.)

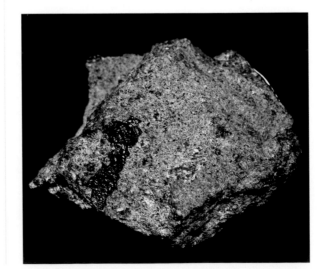

Most basaltic meteorites (eucrites), such as Juvinas which fell in France in 1821, are made up of fragments of igneous rock broken by impacts on the surface of an asteroid. Note the small patch of shiny black fusion crust on the Juvinas fragment.

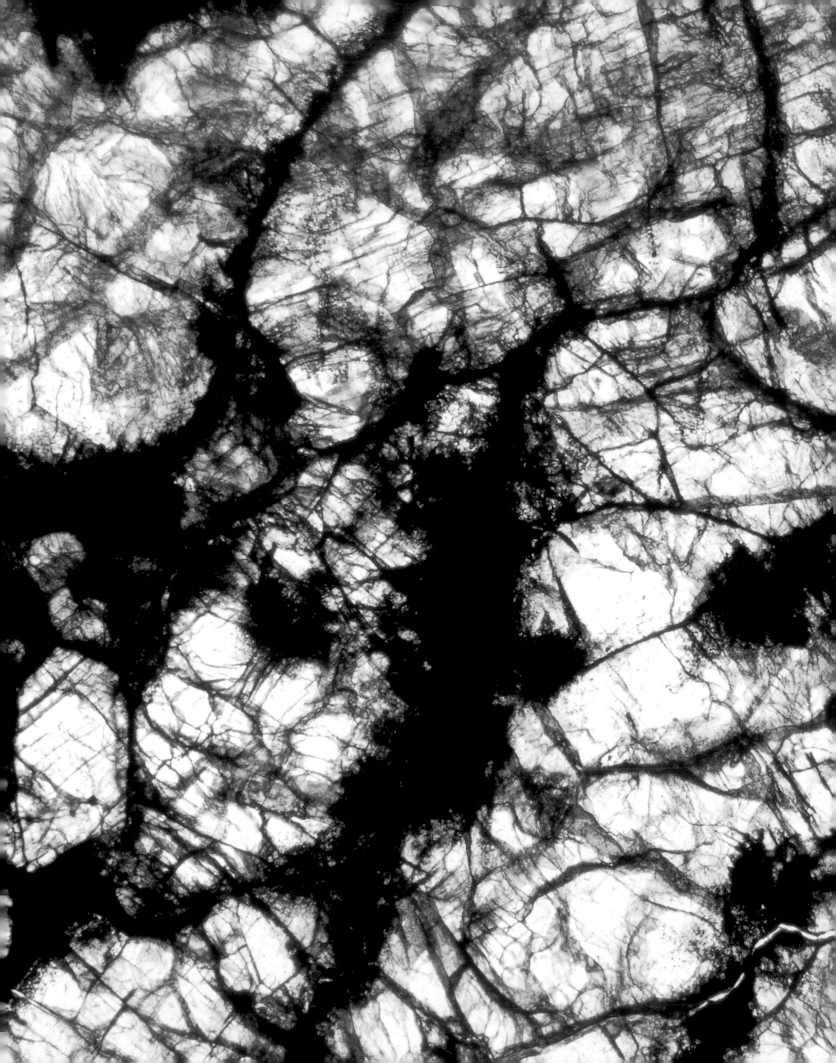

The North Haig ureilite achondrite is an accumulation of crystals of olivine (light) and pyroxene separated by black carbon-rich graphite- and diamond-bearing material. (View 4 mm.)

The Johnstown achondrite which fell in Colorado in 1924 is one of the rare diogenite meteorites.

usually stopped by thick atmospheres like Earth's, but mineral fragments in howardites are intensely and evenly damaged, suggesting they have been regularly turned. These meteorites, then, must have come from the surface of a body with no atmosphere where repeated impacts have 'gardened' the surface debris.

Ureilites, of which around 90 examples are known, are the most enigmatic of achondrites and their origin is controversial. This rare group is named after the Novo-Urei meteorite that fell in Russia in 1886, one of only five observed falls of its kind. Ureilites are among the strangest rocks known to science, with crystals of olivine (iron-magnesium silicate) and a calcium-bearing pyroxene (pigeonite) set in a black, veined matrix containing carbon, metal and sulfides. Their igneous textures suggest that silicate crystals grew from molten magma and settled slowly as cumulates. But this conventional interpretation of ureilites is not universally accepted. Relationships between family members do not accord with a wholly igneous history, and ureilites may be residues from partial melting. Other ureilite family characteristics, such as their oxygen isotopes, seem to be more in common with carbonaceous chondrites.

Subsequent to their assembly, heating of the ureilites removed some iron from the outer rims of olivine crystals where they are in contact with carbon. The carbon is mainly in the low-pressure form of graphite, but some ureilites also contain tiny grains of two high-pressure forms of carbon, diamond and lonsdaleite, offering a clue to part of their history.

Diamond generally forms at high pressure and in the ureilites this was probably generated by a catastrophic collision. The crushing pressure of the impact may have partially converted graphite to diamond. Evidence for impact in many ureilites lies in mineral grains that are crushed and damaged: a few ureilites are mixtures of rock fragments, some resembling carbonaceous chondrites. Because ureilites are rich in carbon and have other similarities with the C-chondrites, their ancestory may lay there, but the family relationship is several times removed.

Howardites, like the Kapoeta achondrite that fell in the Sudan in 1942, are mixtures of basaltic rock fragments. Part of the Kapoeta meteorite has been darkened by intense bombardment by solar particles. (View 4 cm.)

Of all meteorites, the *aubrites*, sometimes called 'enstatite achondrites', have the oddest appearance and are the most difficult to recognise. Nine examples recovered from outside Antarctica were observed to fall. Like the enstatite chondrites, aubrites are made mainly of the iron-free pyroxene, enstatite (magnesium silicate) and pale crystals of this mineral, commonly the size of walnuts, make up a large part of these rocks. Aubrites, like Cumberland Falls, sometimes contain a little metal, are oxygen-poor, and contain fragments of other meteorite types. Like some other achondrites, they have also been mixed with other materials by collisions.

As in enstatite chondrites, an acute lack of oxygen in aubrites resulted in unusual minerals. In addition to oldhamite (calcium sulfide) also found in enstatite chondrites, in aubrites titanium has bonded with nitrogen to form osbornite (titanium nitride). Aubrites are closely related to the enstatite chondrites from which they may have been made by melting, but the exact nature of the family relationship is unknown. Because they contain very little iron, aubrites have clear, glassy or sometimes cream-coloured fusion crusts. In areas strewn with pale rocks, like the Nullarbor or the Sahara, these meteorites would be easily overlooked.

Excluding the 21 Martian and 19 lunar meteorite fragments (all planetary achondrites), the remaining achondrites and related meteorites from asteroids are few. These are the angrites, brachinites, acapulcoites and lodranites together totalling about 37 meteorites.

Angrites are named for the only observed fall of its type: Angra dos Reis (stone) that landed in Brazil in 1869. Another three meteorites found in Antarctica are, in many ways, similar to Angra dos Reis and are certainly related. The angrites are basaltic igneous rocks that differ chemically, notably their higher relative amounts of calcium and aluminium, from the eucrites. Although their formation is poorly understood, angrites record a complex history of evolution in a highly differentiated parent body.

Brachinites form a small intriguing clan of meteorites and, together with the acapulcoites and lodranites that are related to each other, draw a blurred line between achondrites and chondrites. They are all igneous or metamorphic rocks, but like the ureilites their chemistries, and sometimes their textures, hold a tantalising record of a chondritic pedigree. Brachina, a meteorite found in South Australia in 1974, is made mainly of olivine. On Earth, such rocks are called 'dunites' and for a long time Brachina sat in splendid isolation amongst meteorites. Since 1984, another six brachinites have been found.

The *acapulcoites* take their name from a meteorite that fell near the famous Mexican resort in 1976; the *lodranites* from a meteorite that fell in Pakistan. The mineral make-up of both, although variable, resembles ordinary chondrites. But other features, like their oxygen isotopes, show that they are different from any of the known chondrite groups. A few acapulcoites, like the Monument Draw meteorite found in the USA, also contain ghostly relics of chondrules, confirming their parentage. The importance of these meteorites is that they are partly melted chondrites, offering family snapshots of melting and migration in their planet-like asteroids in the early Solar System.

Top left: The 4.45-kg Mayo Belwa achondrite, which fell in Nigeria in 1974, is an aubrite. Aubrites are made up of large white crystals of the magnesium silicate 'enstatite'. See-through crusts are typical of these meteorites.

A cross-section of the Cumberland Falls aubrite, which fell in Kentucky in 1919, shows that mixing from impacts has introduced black inclusions of chondrite (illuminated to show their metal particles).

The Reid 013 brachinite achondrite, found on the Nullarbor Plain, exhibits a granular mixture of olivine, pyroxene and feldspar. Brachinites are primitive achondrites that still retain a record of their chondritic parentage.

ROCKS FROM THE MOON AND MARS

On 28 June 1911, at about 9 o'clock in the morning, the peace of the district around the sleepy village of El Nakhla El Baharia, near Alexandria in Egypt, was shattered by a series of loud bangs. In a local hamlet a dog pricked up its ears at the strange sounds. Seconds later it lay dead. The meteorite which killed the dog, and around 40 other pieces totalling 40 kg scattered over a wide area, was gathered up. Seventy years later scientists would show that the Nakhla meteorite was not any ordinary meteorite, but is actually a piece of the planet Mars.

Today at least 21 meteorites from 16 different falls are known to have come from the 'red planet'. Six of these Martian meteorites were found in Antarctica where they had lain in the ice for tens of thousands of years. Like Nakhla and Zagami that fell in Nigeria in 1962, another two of the non-Antarctic samples were actually observed to fall: the Shergotty meteorite that fell in India in 1865, and the Chassigny meteorite that fell in France in 1815. Another 11 fragments of six meteorites were found at Governador Valadares (Brazil), Dar al Gani 476/489/670/735 (Libya), Lafayette and Los Angeles 001/002 (USA), and Sayh al Uhaymir 005/008 and Dhofar 019 (Oman). Although all came from Mars, the initials of just three of these meteorites (SNC) are commonly used to name this rare group.

How do we know that the meteorites came from Mars? There are several key bits of evidence, leaving no doubt. Most Martian meteorites, which are all igneous rocks, formed more recently than other meteorites. Some formed as little as 330 million years ago, suggesting that they came from a body that was large enough and hot enough to have had recent volcanic activity.

So far, NASA's many Mars missions have included the two Viking unmanned probes launched in the mid 1970s, the more recent Mars Pathfinder with its little robotic vehicle Sojourner, and the orbiting Mars Global Surveyor. The Viking probes analysed soil and atmosphere samples and relayed the results to Earth. The proportions of carbon dioxide, nitrogen and the rare gases xenon, krypton, neon and argon trapped inside some of the SNC meteorites are identical to those of the Martian atmosphere measured by the Viking probes. The fact that, among meteorite collections, there are also 19 pieces of the Moon from perhaps 14 different falls shows that rocks can be blasted off another large body to land on Earth. Mars' gravitational pull is only a third of Earth's so it would be easier for fragments to escape.

In the Antarctic summer of 1979–80, Japanese meteorite hunters searching the icesheet at Yamato found a small 52.4-g stone. The 1197th meteorite found that year, its unusual bubbly brown crust and interior of jumbled grey and white fragments set it aside from most other stones. Later its familiar chemical and mineral make-up betrayed its origin as being the Moon.

Until recently, all the lunar meteorites had been found in Antarctica. The first from outside Antarctica, Calcalong Creek, was found in the outback of Western Australia in 1991. A further two samples, Dar al Gani 262 and Dar al Gani 400 have since been found in the Libyan Sahara. The significance of the lunar meteorites is that they come from areas of the Moon not sampled by manned or un-manned missions.

How did the Martian and lunar meteorites get here? Grazing impacts from asteroids or comets on their surfaces probably launched them into Earth-crossing orbits. Extensive impact damage to some of the Martian meteorites is consistent with their violent ejection from the planet's surface.

The Nakhla Martian achondrite, which fell in Egypt in 1911, caused one of only two recorded animal fatalities from meteorite falls: that of a dog.

The lunar meteorite Yamato 791197 exhibits angular fragments of feldspar (white) mixed with other rock fragments.

Mars under the microscope: Nakhla reveals closely interlocking crystals of pyroxene, showing that it crystallised from molten magma. (View about 5 mm.)

The Martian surface as seen from the Viking spacecraft.

less than 1 kilometre t
greater than 10 kilomet

silicate

iron nickel metal

Observe the small facts

upon which larger inferences depend.

SIR ARTHUR CONAN DOYLE

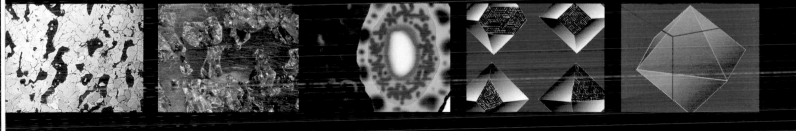

iridium 'rich', and nickel 'rich' iridium 'poor' members. As compelling evidence that most irons are igneous in origin, these trends are exactly those predicted by the fractional crystallisation of molten metallic magma.

Group IAB-IIICD irons lack the steep chemical trend in nickel and iridium shown by the other groups. Irons in group IAB-IIICD show little variation in their gold and iridium contents, although the amounts of nickel they contain range from 6-25 per cent. This lack of distribution behaviour in their trace elements suggests that they were never completely molten. Corroborating evidence that this is the case lies in the inclusions of silicates that occur in many IAB irons. Some of these silicates have the chemical make-up of chondrites, showing that metal never completely separated from its parent material. Some irons, then, had a partly molten birth, perhaps forming pools of metal mixed with stony minerals inside their parent asteroids.

Structures produced by the original solidification of asteroidal cores are rarely seen on slices through average-sized iron meteorites. The reason is that the metallic cores of planet-like asteroids probably

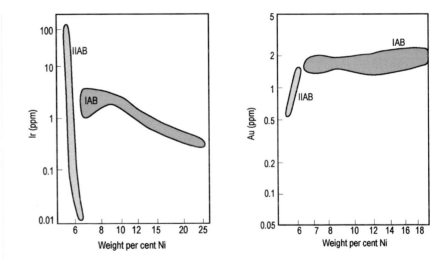

The iridium (Ir) and gold (Au) contents of some iron groups such as IIAB vary greatly from low to high nickel (Ni) members. This is because of the strong separation of iridium and gold between solid and liquid metal. This plots as steep trends for these elements in the so-called 'magmatic' iron meteorite groups that formed in the molten cores of asteroids. By contrast, iridium and gold vary only slightly despite a large variation in nickel content in IAB irons: evidence that these meteorites were never completely molten.

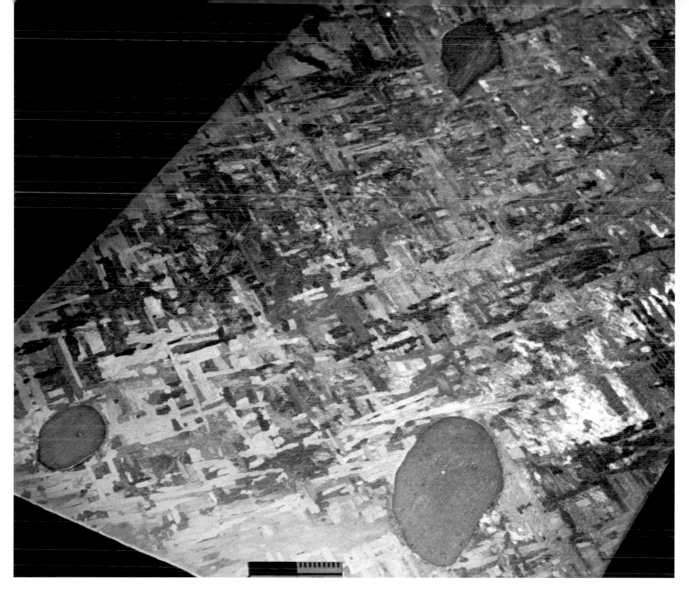

solidified as branching, tree-like crystals called 'dendrites' that were many kilometres long. The distance between dendrites would be around 500 m, very much larger than the size of recovered meteorites. But the intervening areas between large dendrites would have finer branches. Now and then, irons show something of their 'cast' structures. For example, two group IAB-IIICD meteorites, Pitts and Mundrabilla, are made of globular dendritic crystals of metal that have trapped troilite.

In the same way that ice forms across a pond, metal solidifying as dendritic crystals does so directionally. Distributions of elements in irons are consistent with the growth of large elongated crystals in their parent cores. Although we do not see large crystals, stretched troilite nodules seen trapped in the metal of the Cape York group IIIAB iron favour at least one direction prevailing during solidification.

Around 100 analysed irons do not belong to any of the chemical groups. Some of these ungrouped irons have features in common with each other or particular groups, while others are unique. If the ungrouped irons represent samples from different parent bodies then upwards of 50 parent asteroids may be represented by the irons as a whole. Although the precise nature of the parent asteroids of irons is unknown, something can be deduced about their size from their crystalline structures.

Cooling off

Irons frequently consist of three iron-nickel minerals — one low in nickel, called kamacite; and two high in nickel, called taenite and tetrataenite — arranged in a regular trellis-work structure of interlocking crystal plates. Cut surfaces of irons polished and treated with acid reveal this beautiful mosaic. Called a Widmanstätten pattern, the structure is named after the Austrian aristocrat who was one of the first to describe it.

On a microscopic scale, Widmanstätten structures can been made in laboratories, and knowledge of how the structure forms is extensive. Both kamacite and taenite are cubic minerals: that is to say their regular internal stacking of iron and nickel atoms follows the direction of a cube, with planes of atoms intersecting at right angles.

At temperatures above 750°C, solid iron-nickel metal with around 10 per cent nickel forms solid crystals of taenite. On cooling very slowly, plates of kamacite form and grow along regular diagonal planes within taenite crystals. Four pairs of parallel plates of kamacite cut across the corners of the original taenite cube making an eight-faced, diamond-shaped structure called an octahedron. Traditionally, irons that show Widmanstätten structures are called octahedrites.

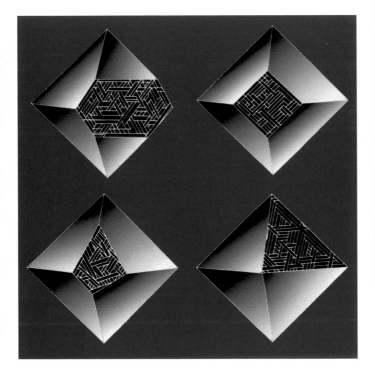

Kamacite grows by the movement of nickel atoms through hot but cooling metal, and both kamacite and its host taenite increase in nickel content. This is achieved by the growth of plates of kamacite at the expense of taenite. With falling temperature, the movement of nickel atoms becomes sluggish. Nickel rejected by the growing kamacite builds up in residual taenite. In the last stages of cooling, nickel diffusion in the kamacite plates slows and a small reduction of nickel content develops in the kamacite in contact with taenite. If cooling is slow, the nickel content at the margins of residual taenite rises dramatically. If slow cooling is maintained and the nickel content rises above 45 per cent, a new mineral (tetrataenite) forms, with a different crystal structure.

In laboratory experiments with iron-nickel metal, it takes months to grow plates of kamacite that are only a few thousandths of a millimetre thick.

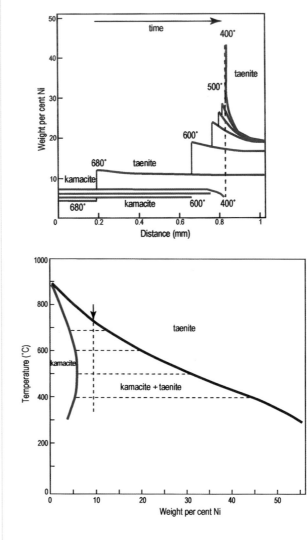

The idealised development of nickel distribution between kamacite and taenite during cooling and growth of the Widmanstätten pattern.

The iron-nickel phase diagram relates temperature and nickel content for cooling iron-nickel alloys.

Far left: The appearance of the Widmanstätten pattern depends on the angle of cross-section through the octahedron.

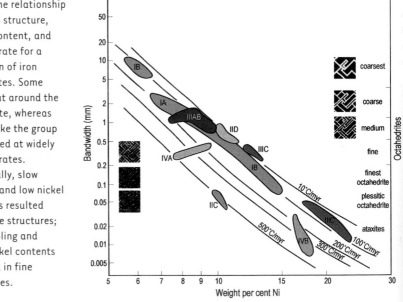

Right: The relationship between structure, nickel content, and cooling rate for a selection of iron meteorites. Some cooled at around the same rate, whereas others like the group IVA cooled at widely varying rates. Essentially, slow cooling and low nickel contents resulted in coarse structures; fast cooling and high nickel contents resulted in fine structures.

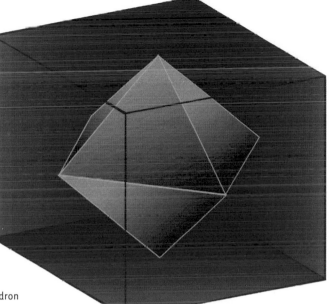

The relationship between an octahedron (blue) and a cube (red). In octahedrite iron meteorites, the plates of the mineral kamacite grow parallel to the faces of an octahedron.

In some nickel-rich meteorites, such as Warburton Range, kamacite is only visible under the microscope as fine spindles.

What is interesting about meteoritic irons is the scale of their Widmanstätten structures, that are often visible to the naked eye and may contain plates of kamacite ranging up to 1-2 cm thick. Clearly, those structures formed over a long time as the result of extremely slow cooling. Crushing pressures would affect the resultant chemistries of these metallic minerals, but the measured compositions of kamacite and taenite in irons closely match those determined in the laboratory at surface pressures on Earth. This is the first clue to suggest that their parent objects were not very large.

Two factors which control the width of kamacite plates in the Widmanstätten structure are the nickel content of the original taenite crystals, and the rate the metal cooled. The lower the nickel content of irons, the higher the temperature at which kamacite plates started to form.

Alternatively, the faster the metal cooled, the less time kamacite plates had to grow. Generally, high nickel contents and faster cooling times result in finer Widmanstätten structures. The scale of the resulting structures reflects a balance between the overall content of nickel and the rate of cooling of the metal.

Iron meteorites with greater than 15 per cent nickel developed Widmanstätten structures at low temperatures. Irons containing microscopic spindles of kamacite are called *ataxites* (from Greek meaning 'without structure'). Irons with less than 6 per cent nickel show no Widmanstätten structures, and are made almost entirely of kamacite. These are called hexahedrites (Greek *hexahedron* meaning 'cube', which has six sides). Meteorites with the hexahedrite structure belong to the IIA end of chemical group IIAB.

Since the average nickel contents of meteorites and the widths of their kamacite plates can be measured, we can calculate the rate at which Widmanstätten structures formed, which reveals something about the size and nature of their parent asteroids. For group IIAB irons that lack Widmanstätten structures, cooling rates can be estimated from the growth of other minerals, such as schreibersite (iron nickel phosphide).

Computer models used to simulate the growth of Widmanstätten structures, so reproducing the nickel concentration profiles in kamacite and taenite, estimate that many irons cooled at rates of around a few degrees every 100 000 years. Widmanstätten structures thus took millions of years to form, while cooling from 600°C to 400°C. This could only have happened if the materials we now see as iron meteorites were once thermally insulated by deep burial inside their parent bodies.

Strike while the iron is hot!

Consistent with an origin from a single asteroidal core, the rates at which group IIIAB irons cooled cluster around a narrow range of few degrees to a few tens of degrees per million years. In contrast, despite showing a strong chemical trend indicating they crystallised from a single body of metallic melt, group IVA irons have widely differing cooling rates: from a few tens of degrees to several hundred degrees per million years. To explain this inconsistency some scientists suggest that, while it was still hot, the parent body of IVA irons was blasted to pieces by

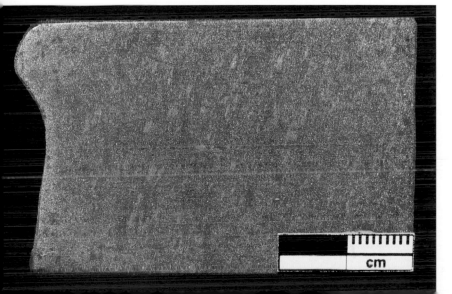

cm

a collision. Unable to escape the gravitational pull of the fragments around them, the pieces of the parent later re-aggregated to form a new 'rubble pile' asteroid with raisins of metal cooling at different rates.

From the measured rates at which irons cooled, and assumptions about the insulating properties of the silicates that once mantled them, calculations show that their parent bodies ranged from a few tens to a few hundreds of kilometres in diameter, reassuringly consistent with the observed sizes of asteroids.

As to how metre-sized chunks of metal excavated from hundreds of kilometres depth in their parent asteroids to arrive on Earth as meteorites there is only one plausible explanation: catastrophic destruction from collisions. But what evidence is there to support this? Under the microscope, an enormous range of alterations caused by high-pressure impacts are found in iron meteorites, revealing everything from mildly deformed metal, through severe alteration by heat generated during violent impacts, to remelting of once slowly cooled metal.

Mix and match: the IIE irons

Explanations for many of the features of meteorites are rarely proven beyond doubt, so for as long as they fit the facts, they remain 'theories'. Established theories may be rendered unworkable by new facts, which then have either to be modified or discarded completely. New, more robust, theories are constructed to accommodate the facts, and in a 'two steps forward, one step back' fashion the truth is gradually approached. Not surprisingly, because of the nature of the study, meteoritics is littered with splendidly complicated examples.

Meteorites from a single family, but with considerable, sometimes conflicting individual variations, lend themselves to equally diverse explanations of their origin. The great majority of irons appear to be fragments broken from asteroidal cores where (pallasites aside) extensive melting caused near perfect separation of metal from silicates. Irons with a variety of relic silicates have more complex histories.

Amongst the most intriguing of irons with silicates are group IIE. Out of the 18 or so family members, eight are raisin-bread mixtures of metal and silicate with a variety of textures. Centimetre-sized globules of silicates in Australia's Miles and in Colomera from Spain, contrast with a large, tongue-shaped lump of stone swathed in the metal of the Nullarbor find, Watson.

Barely discernible chondrules, typical of ordinary chondrites, in the stony parts of the Russian IIE iron, Netschaëvo, pinpoint its parentage. Watson's stony portion also has the chemistry, without metal and sulfide, of an H-group chondrite, but silicates in other IIE irons, like Weekeroo Station, are unlike any chondritic meteorite. The overall oxygen isotopic mix of the stony portions of IIE irons is similar, but not identical, to H-chondrites, suggesting they may be distant cousins rather than true family.

Often made-up of many small original crystals of taenite, slow cooling having provided each with its own Widmanstätten structure, the metal of IIE irons with silicates is heavily kneaded and beaten. The abundant parental taenite crystals grew at high temperatures, their damaged condition recording later collisions. In another Russian iron, Verkhne Dnieprovsk, metal was hit so hard that seed-sized crystals of schreibersite (iron nickel phosphide) 'popped' instantaneously into globules of melt.

Two very different theories suggest that the IIE irons formed either from pools of impact melt on the surface of an asteroid; or, much like the IAB-IIICD irons, by part-melting deep within their parent body. Although some trace element trends in IIE irons do not point to a single core, they support a partly molten origin.

Contradictory evidence comes from trace elements, like gold and nickel, showing trends typical of fractional crystallisation. Consistent with deep burial, unaltered IIE metal apparently cooled at various, but generally slow, rates (from 1–100°C/million years). A more complicated variation on the two opposing theories proposes the IIE irons initially solidified from a single pool of metal, followed by a history of impact

Above: Asteroids with a 'rubble-pile' internal structure are possibly created by the re-aggregation of debris from the destruction by impact of a differentiated asteroid. In this way the previous thermal history of the original asteroid is overprinted by the later event.

Left: Polished and etched with acid, a cross-section of the Mount Edith group IIIAB medium octahedrite shows kamacite plates less than 1 mm thick, laths of schreibersite (iron nickel phosphide) and nodules of troilite.

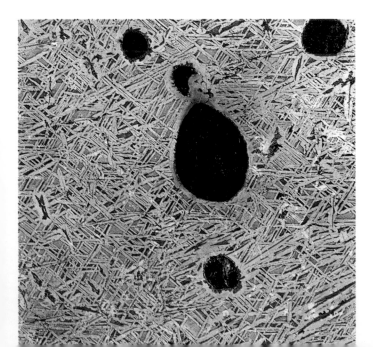

Right: Mechanical shock twins, called 'Neumann bands' (fine lines), in the metal of an iron meteorite formed by metal fatigue, caused by collisions in space. (Magnification x 100.)

Right: Mesosiderites, such as Mount Padbury found in Australia in 1964, are complex mixtures of metal (bright) and basaltic rock fragments (black) of diverse origins brought together by shock mixing.

melting, dispersal, mixing with silicates, and finally asteroidal burial at different, but considerable depths.

The stony portions of IIE irons vary from being 'primitive', that is close to their chondrite parents, to highly 'differentiated' or greatly removed from their original composition. But the extent of the evolution of their silicates does not correspond with increased nickel contents resulting from fractional crystallisation, suggesting that a single melting process was not responsible for the features of IIE irons. Similarly, no parallel exists between cooling rates and the nickel composition of IIE metal.

Complex rocks, with histories of more than one episode of melting and mixing, IIE irons remain enigmatic. Recent identification in the Main Belt of asteroids of a possible parent for this unusual group may also account for the elusive source of the H-chondrites, adding another dimension to the IIE story.

About 200 km in diameter and long suspected as a potential contributor of meteorites, asteroid 6 Hebe orbits close to regions of instability with Jupiter that could have launched detached fragments into Earth-crossing orbit over millions of years.

Analysis of light from Hebe's surface suggests a make-up similar to H-group chondrites, but does not exclude other meteoritic rocks. Spectral variations across the asteroid's surface may mark deep craters and indicate

metal-rich pods. If Hebe is the co-parent of H-group chondrites and IIE irons, or something like them, several ingenious ways have been put forward to account for the unusual surface features of the asteroid.

Large impacts into metal-rich H-chondrite rock might have produced giant pools of melt that sepa-

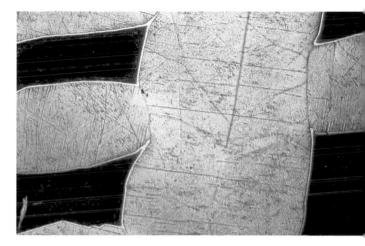

rated into thin layers of dense metal covered by a float of lighter silicate. Haphazardly, later impacts shattered, melted and mixed the resolidified layers producing a veritable rock garden of metal-rich and silicate boulders.

Alternatively, the metallic melts may have been splashed from a large iron projectile, perhaps a sizeable core fragment from another source, that collided with Hebe. Opponents to an asteroidal surface origin for the IIE irons cite the inefficiency of impact as a way of melting large volumes of rock. In support of an impact origin, igneous fragments found in many H-group ordinary chondrites are typical of a rapidly melted chondrite drained of metal and sulfide.

Some IIE irons with silicates appear to be only 3700 million years old, significantly younger than other meteorites. While there is some argument as to what their 'ages' actually mean, if it is the time when they formed then it is difficult to explain how, other than by impact, they were melted — since most meteoritic parents appear to have been thermally 'dead' by that time.

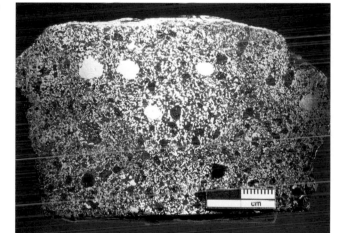

The Miles group IIE iron meteorite, found in Queensland, is a complex mixture of metal and silicate (dark patches).

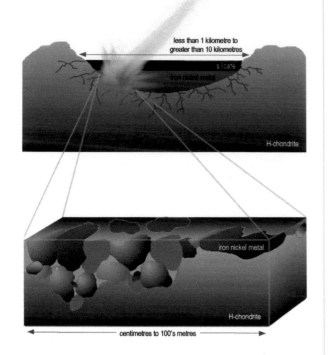

In favour of the deep burial of IIE irons are their apparently slow cooling rates. But what if these are wrong? The 'core' theory would be difficult to support. Because of the suggested link between IIE irons and H-chondrites, there is an air of circularity about the arguments that link them to asteroid Hebe, but the similarity of the deduced history of IIE irons and the observed features of Hebe beg the question: if they are not from Hebe, where are Hebe's meteorites?

When faced with two or more equally plausible theories, scientists often invoke Occam's Razor. Named after William of Occam, a fourteenth century theologian who played a pivotal role in bringing logic and science into the post-mediaeval world, Occam's principle stresses that entities must not be multiplied beyond what is necessary. Essentially, provided it is consistent with the facts, the simplest theory should be favoured over unnecessarily complicated explanations. In much the same way as governments 'guillotine' debates, the theory favoured by Occam's Razor holds until it is shown to be wrong. For the IIE irons, whichever way we look at it, even the simplest explanation is complex.

Stony-irons

Stony-irons are by far the rarest of meteorites, totalling only 116 known examples. The pallasites and mesosiderites, two unrelated families of silicate and metal mixtures, make up the stony-irons. The *pallasites*, of which about 50 are known, are made of crystals of olivine and sometimes pyroxene set in metallic iron-nickel that sometimes has a

Widmanstätten structure. The *mesosiderites* are made up of mixtures of irregular and rounded blocks of rock, similar to some achondrites in composition, with nuggets and veins of iron-nickel metal. Metal nuggets in mesosiderites can also show Widmanstätten structures.

The most plausible explanation for the formation of stony-irons is by the mixing of both solid and liquid metal and silicates at various depths within their parent planets. A close chemical relationship between more than 40 pallasites and at least one family of irons (group IIIAB), suggests that they formed in the region between the metallic core and rocky mantle of the same small planet-like asteroid. Mesosiderites, on the other hand, represent impact mixtures of solid and liquid metal and achondritic rocks of diverse origins.

If there was ever a meteoritic family of mongrels it is the mesosiderites. These complex rocks, of which only about 60 are known, probably resulted from a series of events in the early Solar System. Mesosiderites hold clues to the evolution of some asteroids within a short time of their formation, painting a picture of melting, catastrophic collisions and re-aggregation at various times in their history.

Overall, the silicate portions of mesosiderites resemble howardite achondrites. Igneous silicates in mesosiderites in general resemble the howardite/eucrite/diogenite clan of achondrites, and it has been suggested that they come from the same parent asteroid.

The mysteriously missing mantle meteorites

In science it is not always the evidence presenting itself that is important. Something that should have happened but is missing can be just as informative. The lack of meteorites with the characteristics of mantle rocks is one of meteoritics' great unsolved mysteries.

Complete melting and separation of an asteroid with a diameter of about 200 km should result in a metallic core about 100 km across with a mantle of olivine-rich rocks about 45 km thick topped with a 5-km thick crust of basaltic rocks. Iron meteorites are abundant: solid evidence that some asteroids melted and differentiated. Their cooling rates implicate parental asteroids with dimensions in the range of hundreds of kilometres. Complementary achondrites are also relatively numerous among meteorites. What is missing are achondrites rich in olivine that

One way that the metal of group IIE irons may have formed and later mixed with H-group chondrite silicates is impact gardening on the surface of asteroid 6 Hebe.

Idealised view of a differentiated asteroid. In the meteorite record there are iron meteorites from the cores of differentiated asteroids, and abundant igneous achondrites that came from the crusts. However, olivine-rich meteorites from the mantles of such asteroids which should be numerous are virtually unknown.

A slice of the Imilac pallasite stony-iron, found in Chile in 1822, shows bright metal enclosing green olivine. Pallasites probably came from the regions between the metallic cores and olivine-rich mantles of differentiated asteroids.

should be more abundant than irons and basalt-like achondrites put together. Like Earth, around 70 per cent of a differentiated asteroid should be taken up by olivine-rich mantle rocks.

A few asteroidal olivine-rich meteorites, such as Brachina and its relatives, appear possibilities — but their chondritic chemistries rule out a highly differentiated origin. Ureilites might be from mantle regions of their parent body, although their make-up suggests that it had only partly melted. While pallasites are prime candidates for core-mantle boundary rocks, mesosiderites, with their complex history of asteroidal break-up and mixing, should also contain abundant olivine — but it rarely makes more than a couple of per cent of their volume.

So where have all the olivine-rich rocks gone? Were they simply not knocked into Earth-crossing orbits, or is there some other explanation? Differentiated meteorites were made from original chondritic materials with wide ranges of compositions, perhaps even more diverse than chondrites preserved in collections today. The general absence of mantle meteorites might be that among the asteroids, melting and separation took place from a variety of starting materials, in a wide range of environments and under the influence of numerous forces, not all of which led to abundant olivine-rich rocks.

Those achondrites retaining close chemical similarities with their chondrite parents show that the amount of melting experienced by certain asteroids varied greatly. Others, like the eucrites, diogenites and howardites, and the group IIIAB irons, must have come from bodies that melted extensively, and therefore should have associated olivine-rich rocks.

It is likely that the answer lies in the way asteroids were broken up. Ages determined by exposure to cosmic rays for iron meteorites originating from cores show they orbited in the Solar System for around 400-600 million years before landing on Earth. Exposure ages for irons range from 200-1000 million years, up to 50 times longer than some stones. Similar exposure ages for all members within groups such as IIIAB and IVA point to the destruction of their parent bodies in single, catastrophic events. Because they are weaker than irons, during a long period in space fragments of asteroid mantle may have been ground away slowly, the remnant dust spiralling towards the Sun

Further meteoritic mysteries are the apparent absence of basalt-like complements to the aubrites (enstatite achondrites) and ureilites, since the partial melting of any known chondrite should produce a substantial amount of basalt-like rock. The whereabouts of those basalts that probably formed crusts on the asteroids from which the magmatic irons came are also unknown. A fate similar to the olivine-rich rocks is one possibility for these missing meteorites. Another theory is that they were blown away as pyroclasts from their parent asteroids during violent volcanic eruptions.

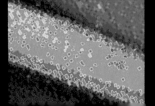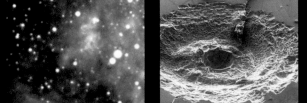

Twinkle, twinkle, little star,

how I wonder what you are!

Up above the world so high,

like a diamond in the sky!

JANE TAYLOR, 1805

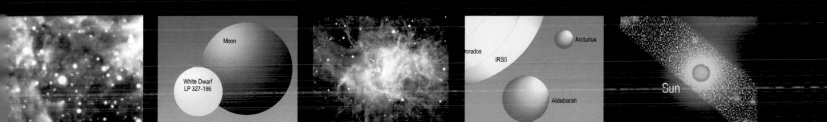

STAR DUST

Along the road to cosmic enlightenment, many meteoritic discoveries have challenged scientists to move in new directions. Sometimes the driving force came from falls of rare meteorites. Always the challenge is to delve more deeply into their interiors, isolating smaller and smaller grains. Now and then, discoveries are made that revolutionise our understanding of the Solar System. One of the outstanding milestones of the twentieth century was the discovery that chondrites contain tiny grains from *beyond* the Solar System.

Half a century ago, a physicist's view of the Universe was elegantly simple. All but the lightest elements had been made during the birth of the Universe with uniform isotopic abundances. In the early 1950s, seemingly in support of this view, measurements had not revealed any differences between the abundances of isotopes in meteorites and the Earth. By the late 1950s, when it was understood that stars are nuclear atomic factories, the picture changed completely.

In 1957, Margaret and Geoffrey Burbidge, together with William Fowler (in California) and Fred Hoyle (then at Cambridge), drew up a blueprint of nuclear reactions in stars that would produce all but a few of the elements and isotopes. Alastair Cameron of the Atomic Energy Commission of Canada came to similar conclusions independently. The Universe may be around 12-15 billion years old. Since shortly after the Big Bang heralded the birth of the Universe, stars have manufactured isotopes of the chemical elements. The enormous pressures and temperatures at the centres of stars, like our Sun, cause atoms of hydrogen, the most abundant element in the Universe, to fuse into atoms of helium. By 'burning' helium and through other nuclear reactions, stars create, atom by atom, isotopes up to heavy iron. Massive stars with 'burnt-out' iron cores become unstable, eventually exploding as supernovae, spilling their cargo of newly-formed isotopes into space. New generations of stars form and the process continues. Through the life and death of stars, isotopes of the elements are made.

In this very different view of the Universe, the isotopic mixes of elements should vary in space and time, but conveniently carry the designer labels of their manufacturers — the stars, supernovae or novae in which they were made. By the 1960s the search for tell-tale isotopic anomalies in meteorites had started in earnest. The task was daunting but the prize would be great, for finding pristine star dust would reveal something of the environment in which the Solar System formed.

A whiff of suspicion that the infant solar cloud had been bombarded by atomic debris from a nearby supernova was aroused in 1960. A small amount of a rare gas, xenon, extracted from the Richardton ordinary chondrite, which fell in the USA in 1918, was found to contain more of the isotope xenon-129 than predicted. The excess was attributed to the decay of the now extinct radioactive isotope iodine-129 with a half-life of only 16 million years. Iodine-129 was made in a supernova shortly before, cosmologically speaking, it was incorporated into the rock. Because the 'daughter' xenon gas was not lost to space, the material must have cooled rapidly. For the record to have survived, no more than ten half-lives (less than 170 million years) could have passed between the manufacture of iodine-129, and the birth of the Solar System 4560 million years ago.

Unusual xenon was also found in the Renazzo carbonaceous chondrite in 1964. Excesses of other

Tiny diamonds measuring only around 40 millionths of a millimetre, found in the Allende meteorite which fell in Mexico in 1969, probably formed in the gas streaming from another star. These grains are older than our Sun and planets.

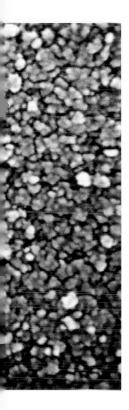

Right: The white inclusions visible in the Allende carbonaceous chondrite formed at high temperatures at the beginning of the Solar System. The mixture of oxygen isotopes that they contain originated beyond our Solar System.

xenon isotopes (131 to 136) were measured in the Pasamonte eucrite achondrite in 1965. These were attributed to the decay of now extinct plutonium-244 with a half life of 82 million years — confirmation that young material had been added to the early Solar System. Solid pieces of this material need no longer exist, for even if the original grains were destroyed, as in the case of the melting that produced the igneous Pasamonte meteorite, any remaining radioactive isotopes would leave a fingerprint of their former presence in their decay products.

The vintage year for planetary science was 1969. On 20 July a man first set foot on the Moon, but at either end of the same year and on opposite sides of Earth two meteorite falls had a more profound effect on our understanding of the Solar System. In the early hours of 8 February, more than two tons of fragments of a CV3 carbonaceous chondrite fell over a large area around the small Mexican town of Pueblito de Allende. Seven months later, on 28 September about half a ton of fragments of a CM2 carbonaceous chondrite pelted the Australian country town of Murchison, in Victoria Of kinds previously known in meagre amounts, the Murchison and Allende meteorites provided a wealth of material on which to work.

In 1969, scientists also extracted neon gas from carbonaceous chondrites. By heating small samples, neon was trapped and analysed in a mass spectrometer. Expecting to find the three isotopes of neon (neon-20, neon-21 and neon-22) in the same proportions found elsewhere in the Solar System, they were surprised by a tiny portion, rich in neon-22, that was very different from anything found before. This kind of neon (later called 'neon-E') results mainly from the decay of sodium-22, produced during novae. Sodium-22 has a half-life of only 2.6 years so it must have been trapped in solid grains very quickly.

Over the next two decades a small army of scientists probed deeply into the Murchison, Allende and other carbonaceous chondrites. It was not long before more strange isotopic mixtures were found. In 1973, oxygen isotopes in the white, high-temperature calcium-aluminium-rich inclusions of Allende were found to be mixed in very different proportions to bodies like the Earth. By blending with other materials, non-radioactive or stable isotopes, such as oxygen, can lose their identities very easily. The inescapable conclusion was that the unusual oxygen must be a relic from beyond the Solar System. Of the greatest significance, the discovery showed that the

The Crab Nebula is
one of the most
prominent remnants
of a supernova. Its
formation was
observed by Chinese
astronomers on 4
July 1054 AD, when it
shone four times
brighter than Venus.

turmoil of the birth of the Solar System had not stirred the elements into an even isotopic mix. Isotopic anomalies of magnesium, silicon, calcium and titanium were also found, although any material in which they might have once resided had long since been destroyed.

It was very unlikely that inert gases, such as xenon and neon, would form compounds of their own, so they had to be trapped in materials of other compositions. Speculation mounted as to what the minerals carrying these unusual isotopic mixes might be. If mineral grains from other stars existed, then they had to be tough enough to have lasted for longer than the age of the Solar System. Since they had not been found easily, then they must be very small.

Neon-E lit the way in the quest to find grains of star dust. In 1974, scientists looking for the neon anomaly in carbonaceous chondrites narrowed it down to two materials: one rich in carbon that released neon at low temperatures; and a denser material that released it at higher temperatures. Another rare gas, krypton, trapped in these materials also showed unusual mixtures of isotopes that could not be explained by a Solar System origin.

Intuitively, it was realised that the best place to look for pristine star dust was in those chondrites that had suffered least alteration. Since they are made of nearly all matrix, the CI chondrites came under the spotlight. Frustratingly, the search became bogged in the gooey organic mess pervading the fine-grained portions of these meteorites that defied the separation of tiny grains. A period of painstaking trial and error to isolate grains with isotopic anomalies ensued.

In 1987, more than twenty years after the discovery of their trapped gas anomalies, the first presolar grains were identified in the Allende meteorite: crystals of diamond only a few millionths of a millimetre across. These interstellar diamonds were the carriers of xenon-HL, a strange mixture of both heavy (H) and light (L) isotopes.

Laboratories across the USA and Europe intensified the search for the carriers of another rare isotopic mix of xenon (xenon-S), and the elusive neon-E. Xenon-S was particularly abundant in the Murchison meteorite, where it appeared to be trapped in a carbon-bearing mineral. Shortly after the discovery of interstellar diamonds, the mineral was shown to be silicon carbide, similar to a tough synthetic material commonly used as an abrasive on Earth.

Further work showed that not one, but two forms of neon-E were trapped in different materials: the low temperature form in graphite and graphite-like carbon; and the high temperature form in silicon carbide. These grains, like the silicon carbide that carries xenon-S, are rich in carbon-13, tens of times the usual amount for the Solar System.

Today, the inventory of star dust includes crystals of oxides, such as corundum (aluminium oxide) and spinel (magnesium aluminium oxide), silicon nitride, as well as diamond, silicon carbide, graphite, and other forms of carbon. Locked away inside some grains of silicon carbide are cores of titanium carbide. Other new discoveries include carbides containing zirconium and molybdenum, iron carbide, kamacite, titanium oxide and hibonite. Each of these ancient grains, some of which appear to be a billion years older than the Solar System, offer a glimpse into the workings of stars.

Star gazing

Uncovering a constellation of grains of star dust in meteorites raised more questions than answers. How are these unusual isotopic mixes to be interpreted? What do they tell us about about the stars in which they formed? Moreover, were the host grains formed at the same time as their anomalous contents and, if so, by what process?

Carbon grains bearing the low-temperature form of neon-E are also rich in nitrogen-15. Novae are the main producers of nitrogen-15 in the galaxy, and the radioactive parent of low-temperature neon-E, sodium-22, is also known to be the product of these events. The relative amount of carbon to oxygen in novae is also high, so a reasonable conclusion is that both the carbon grains and their contents formed in one and the same event.

Just as one might expect, the relationship between most interstellar grains and their contents is not always that simple. Grains of silicon carbide carrying the high-temperature form of neon-E have very variable isotopic mixes of carbon, nitrogen and silicon. This neon-E did not form by the radioactive decay of sodium-22, but was made as a primary product in the helium shell of a type of red giant star. Silicon carbide, it seems, originated in complex ways and the evidence suggests that grains were made at different times and in more than ten different stars.

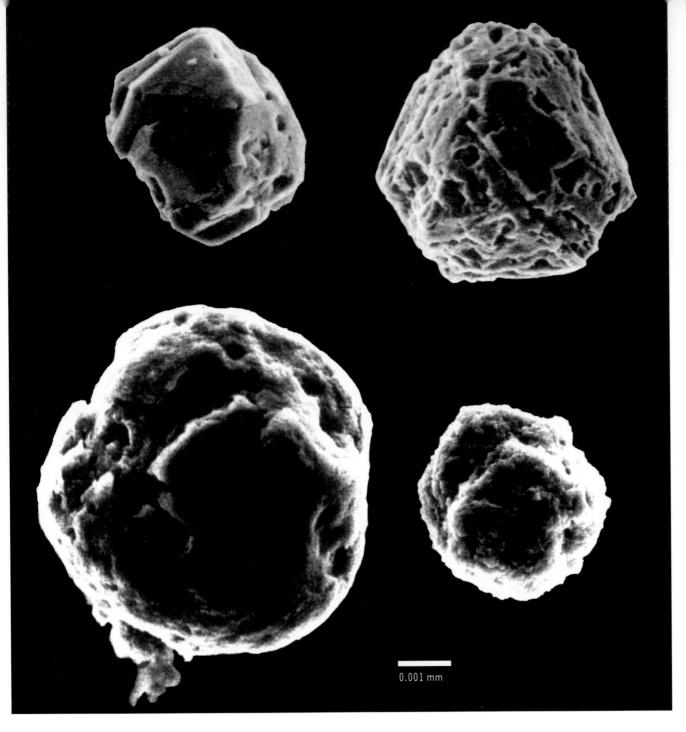

Left: Grains of interstellar silicon carbide (top) may have come from as many as ten stars. *Below:* Presolar grains of graphite.

Diamonds are the most common grains from other stars bound in meteorites, but they make up just four hundredths of one per cent of CM chondrites like Murchison. So small are they that each diamond may only be a few thousand carbon atoms across. No straightforward process is known to produce xenon rich in both its heavy and light isotopes (xenon-HL), so its presence in interstellar diamonds is a mystery. Various beautiful theories have been suggested to account for the unusual mix and the production of diamond, but most founder on ugly little facts.

One suggestion is that diamonds and their trapped gases became mixed during a supernova explosion. Heavy isotopes of xenon are made in the outer carbon shell of a supernova, whereas the light isotopes are made in the hot interior of the dying star. The diamonds might have been made during an earlier phase of the star, and the gases implanted during the explosion. However, stars that generate the same balance of carbon and nitrogen isotopes as in diamonds are red giants, and they do not end their lives in the kind of supernovae that yield xenon-HL.

Collisions between carbon grains hit by the shock wave of a supernova might convert some of them to diamond, trapping xenon gas at the same time. Chance encounters between grains is an inefficient mechanism, and theory predicts that there should be some grains that are mixtures of diamond and unconverted carbon. No such mixtures have been found among interstellar diamonds, suggesting

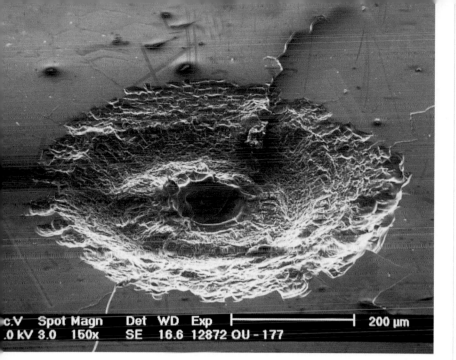

c.V   Spot Magn   Det   WD   Exp |⸺⸺⸺⸺⸺⸺⸺⸺| 200 µm
.0 kV   3.0   150x   SE   16.6   12872 OU - 177

Above: Spacecraft are often involuntary dust collectors. This small crater, only 0.5 mm across, was made on the solar panel of the Hubble Space Telescope by the impact of a tiny dust particle. Some dust enters the Solar System from well beyond it, but such interstellar grains are extremely rare.

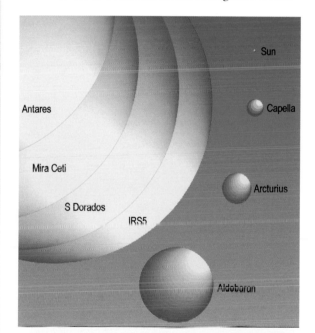

Right: Our Sun is quite small compared to some other stars. Colossal IRS5 is more than 10 000 times the diameter of the Sun.

Right: One of the smallest stars known, White Dwarf LP 327-186, is estimated to be only half the diameter of the Moon, or about 1600 km across.

that impact was not the way in which they formed.

Ironically, the origin of interstellar diamonds may be by the same process used to make industrial diamonds on Earth. Diamond manufacture by the process known as 'chemical vapour deposition' (CVD) entails mixing a source of carbon, such as methane, with hydrogen on a 'seed' material, such as silicon carbide, which has atoms arranged in a simi-

lar way as diamond. When the gases are superheated and the conditions are correct, small crystals of diamond grow on the seed surface. CVD diamonds form at low pressures, which is an important difference between them and other diamonds. Perfect conditions for the growth of CVD diamonds — low pressures, high temperatures and a concentration of hydrogen relative to carbon — are remarkably similar to those observed in the atmospheres of carbon stars.

Interstellar diamonds were therefore probably made in the gases streaming from a carbon star. The expanding cloud of forming diamonds was hit by a nearby supernova charged with xenon-HL, which was incorporated into the diamonds. Superficially, such a mechanism seems too coincidental to be repeated. However, binary stars (star pairs) are common in the Universe. Perhaps after the larger star of a pair collapsed into a white dwarf its partner reached the red giant stage, 'burning' helium and carbon to become a carbon star. The white dwarf would have fed voraciously on its partner, and first silicon carbide then diamonds were formed in the expanding gas around the carbon star. When the white dwarf reached critical mass it exploded as a supernova, impregnating the diamonds with heavy and light xenon isotopes.

Star dust has brought astrophysics and meteoritics closer together. The former provides the theory to understand nuclear processes in stars, while the latter obtains measurements from star dust that test the theory. In the same year as the discovery of interstellar diamonds in meteorites, astronomers observed the brilliant explosion of a supernova (1987 A) in a near neighbour to our galaxy. Observations of this event confirmed theories of the formation of the elements and the important role played by supernovae.

Star dust has been found in all families of chondrites, but grains are most abundant in carbonaceous chondrites. Pre-solar grains decrease in number as chondrites become more crystalline, presumably because they have been destroyed by metamorphic heat. Nevertheless it seems that all the chondrites received some interstellar material during their formation. Other than pointing to the stellar environment in which it formed, unaltered star dust does not reveal much about the early Solar System, as the particles retain no memory of their arrival. Theirs is a truly pre-solar story of nuclear and chemical evolution of stars. Their destruction in recrystallised chondrites, however, is a record of the temperature history of the parent bodies of meteorites in which they once resided.

With present day techniques it takes around a thousand million interstellar diamonds to provide a sufficient sample to analyse for gases. Until they are exposed by more sensitive detection equipment, many of the secrets of interstellar grains will remain hidden for years to come.

A single grain of silicon carbide, plucked from a small impact crater on a solar cell from the Hubble Space Telescope, may be the only example in captivity of interstellar dust entering the Solar System today as a single particle. Contemporary interstellar dust is apparently much rarer than its fossil equivalents in meteorites.

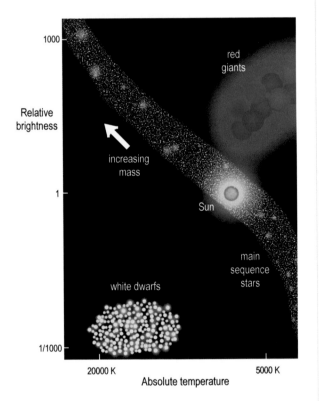

In a star's lifetime it will change in size, temperature and brightness. Stars spend most of their lives in the Main Sequence. The Sun has been in the Main Sequence for 4500 million years, and still has half its time there to go. When the hydrogen in its core is exhausted, the Sun will become a red giant, devouring much of the Solar System. It will then shed its outer material to become a white dwarf. Stars much bigger than the Sun spend less time in the Main Sequence, before exploding as supernovae.

Viewed from Earth,
Supernova 1987A
confirmed theories of
the origin of the ele-
ments.

The cluster of stars
(NGC 4755) known as
the Jewel Box, situat-
ed in the southern
Milky Way, is one of
the youngest known.
The brightest star has
a luminosity of about
80 000 Suns. The
central region of the
cluster is about 25
light years across,
and located near the
centre is a red giant.
The bright stars
will begin their
own evolutionary
expansion in the near
cosmological future.

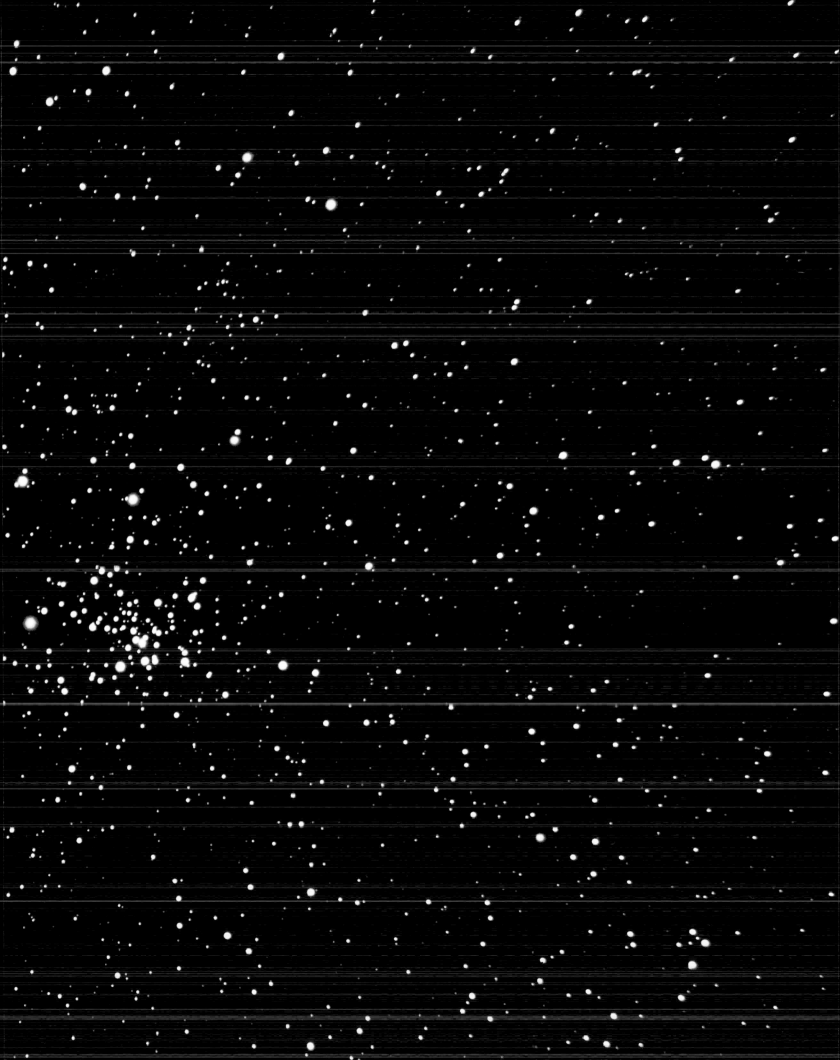

The changing of bodies into light,

and light into bodies,

is very comfortable to the course of nature,

which seems delighted with transmutations.

ISAAC NEWTON

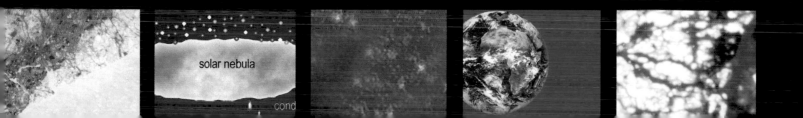

solar nebula

cond

ROCK OF AGES

Although their chemistries and textures reveal much about the ways meteorites formed, including events they were subjected to as rocks, these properties tell us neither when they formed, nor the timing of subsequent episodes in their history. In Chapter 5 we discussed how, in the last century, the age of the Earth was constrained from a knowledge of the decay of radioactive isotopes and the study of meteorites. This estimate of the Earth's age is linked to the reasonable assumption that both meteorites and the Earth formed from similar materials. Most chondrites formed around 4555 million years ago, implying that this is an upper limit to the age of the Earth and a lower limit to the age of the Solar System.

In addition to the age of formation of meteorites, another three important periods of time in their history can be measured. Each casts light on the complex life-time of meteoritic materials from birth in the early Solar System, their cooling in parent objects, their break-up and long sojourn in space, to their arrival on Earth.

The formation interval

The first measurable episode in the history of a meteorite is the time between the manufacture of elements in stars and their later incorporation into the embryonic Solar System. Grains of pre-solar diamond and silicon carbide, hard-won from a wide variety of chondrites, provide a wealth of information on the timing of what is called the 'formation interval', revealing that this period covers a series of separate events spanning more than a billion years before the birth of the Solar System. Each event contributed material tagged with anomalies resulting from now extinct radioactivity.

Among the most recent of these pre-solar events was the formation of a vigorously radioactive isotope of aluminium. Evidence of this lies in the curious white inclusions — made of high-temperature minerals rich in calcium, aluminium and titanium — found predominantly in carbonaceous chondrites.

Shortly after the fall in 1969 of the Allende CV3 chondrite, an unusually high amount of the isotope magnesium-26 was discovered in white inclusions extracted from it. Magnesium has three stable isotopes, abundant magnesium-24, and rarer magnesium-25 and magnesium-26. Magnesium-26 is also the radiogenic daughter product from the decay of aluminium-26, a now extinct radionuclide with a half-life of only 750 000 years. The suspicion was that the measured excess of magnesium-26 in Allende's white inclusions might be a relic of aluminium-26.

In magnesium-rich minerals from the white inclusions, small amounts of radiogenic magnesium-26 would be difficult to detect. Instead, scientists concentrated their investigations on aluminium-rich minerals, such as anorthite (a calcium aluminium silicate), containing only tiny amounts of magnesium as an impurity. Today, aluminium occurs as the stable isotope aluminium-27. If aluminium-26 had been present when anorthite formed in the Allende inclusions, its decay product would make up a significant proportion of the magnesium impurity.

When minerals separated from the white inclusions were analysed, anomalous amounts of magnesium-26 were found to be proportional to the total amount of aluminium and magnesium they contained. The excess magnesium-26 correlates with an increase in aluminium from the inclusions, showing that it is the daughter product of aluminium-26.

Because of its brief half-life, this original aluminium-26 must have been incorporated into the inclusions within a short time after its formation, becoming trapped in the crystallising minerals. Calculation of the relative amounts of radioactive aluminium-26 and stable aluminium-27 originally present in the inclusions reveals that the formation interval was within about 5 million years of the synthesis of radioactive aluminium in a nearby supernova.

Formation ages

The second important 'age' of a meteorite is the time when it was made. For an achondrite or an iron, this is when it crystallised and cooled from molten magma in a parent body. Chondrites are more complex rocks, made up of inclusions, chondrules and matrix. Although they have not been melted since they formed, many chondrites have been altered; ordinary and other chondrites have been heated and crystallised, and some carbonaceous chondrites have been changed by the action of water. Measurements of a variety of different long-lived radioactive isotopes and their daughter products in parts of chondrites reveal a complex history of assembly and processing.

Radioactive 'clocks' capable of measuring great periods of time include the decay of uranium-238 into lead-206; uranium-235 into lead-207; and rubidium-87 into strontium-87. When lead isotopes are measured in whole samples of chondrites they consistently give formation ages of 4555 million years. The same method applied to the white calcium-aluminium rich inclusions in carbonaceous chondrites gives slightly greater ages of around 4560 million years.

The greater antiquity for calcium-aluminium-rich inclusions is supported by the calculated amounts of extinct radionuclides aluminium-26 and iodine-129, both of which appear to have been higher in inclusions than in most chondrules. If the difference can be attributed to the decay of the short-lived radioactive isotopes, then allowing for errors in determining such great ages, the gap in the age of ancient inclusions and younger chondrules in the same meteorite could be as high as 2 million years.

These conclusions are not without question. For example, where were the large, early-formed white inclusions 'stored' before they became part of their chondrite hosts, and how did most escape the apparently later formation of chondrules? But whatever view is taken, the collective evidence favours the theory that the pre-solar cloud evolved into solid material over a surprisingly short time of a few million years.

None of the events affecting chondrites during their residence in asteroids has been resolved satisfactorily in terms of timing. The beginning of alteration on some asteroids, like the alteration of CI chondrites by water, is estimated from the rubidium-strontium isotopic system at around 50 million years

Approximate timing of events (in million years) that led to the formation of the high-temperature calcium-aluminium inclusions (CAI) and chondrules, and their incorporation into rocks on their parent asteroids, and the events which followed.

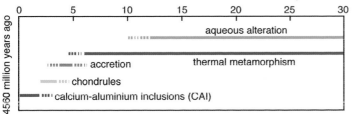

after their formation. Using another radioactive pair (manganese-53/chromium-53) to measure time, alteration may have begun only 20 million years after the formation of high-temperature white inclusions. The iodine-129/xenon-129 system suggests that alteration by water of at least one ordinary chondrite, Semarkona, happened at around 10 million years after the formation of its chondrules.

The onset of the heating which metamorphosed the ordinary chondrites is estimated, from relic radiogenic magnesium-26 in feldspar formed during the process, to within about 6 million years of the formation of calcium-aluminium rich inclusions. Both short- and long-lived radioactive isotopes also support early heating in ordinary chondrites with recrystallisation of their minerals continuing for perhaps tens of millions of years.

Some isotopic systems, particularly those where the decay products are gases like argon, are especially sensitive to heating, and their radioactive clocks are easily 'reset'. They record the high-temperature events to which they were subjected. One such clock measures potassium-40 as it decays to argon-40, with a half-life of 1310 million years. A meteorite's retention or loss of argon gas gives us an indication of its temperature history. Many chondrites' argon-40 ages differ little from ages determined by other methods, showing them to be undisturbed.

A special variation of the potassium-argon dating technique reveals a more detailed picture of the history of some chondrites. Called the argon-39/argon-40 method, the technique requires the conversion of stable potassium-39 to argon-39 by neutron capture in a nuclear reactor. By heating such an irradiated sample to progressively higher temperatures the released argon gas is analysed, and an 'age' for each temperature is calculated from the relative amounts of the two argon isotopes. Argon-39/argon-40 ages of a variety of chondrites vary by around 100 million years either side of a peak at 4500 million years. An explanation of this disturbance is that it is an artifact of the prolonged heating of some chondrite parent bodies.

Long after their formation, numerous chondrites record significant losses of argon, attributed to heating by impacts. Chondrites record two major episodes of argon loss. The first is around 4000 million years ago and corresponds with the heavy bombardment of the Moon. A more recent disturbance at less than 2000 million years ago is thought to reflect the beginning of the break-up of their parent asteroids. Some chondrites, including many L-chondrites, have lost a sig-

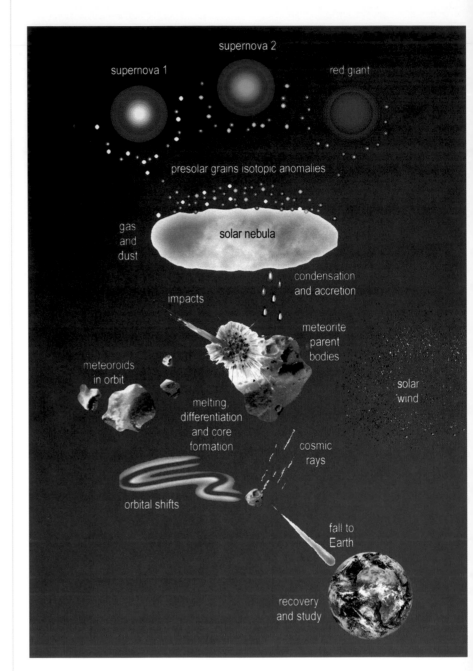

nificant portion of their argon. Many L-chondrites were degassed of argon around 500 million years ago, signalling a major catastrophic event.

Measuring the solidification ages of iron meteorites is altogether more difficult. Because the commonly used radioactive clocks are based on elements within oxygen-bearing minerals, they generally cannot be used to date irons. Other systems of radioactive isotopes can be used, but there is a great deal of uncertainty in the results and their interpretation. Lead isotopes, however, extracted from the nodules of troilite (iron sulfide) in the group IAB iron

The series of events that led to the formation of Solar System materials as deduced from meteorites, their journey through space, and their time on Earth.

CHONDRITES

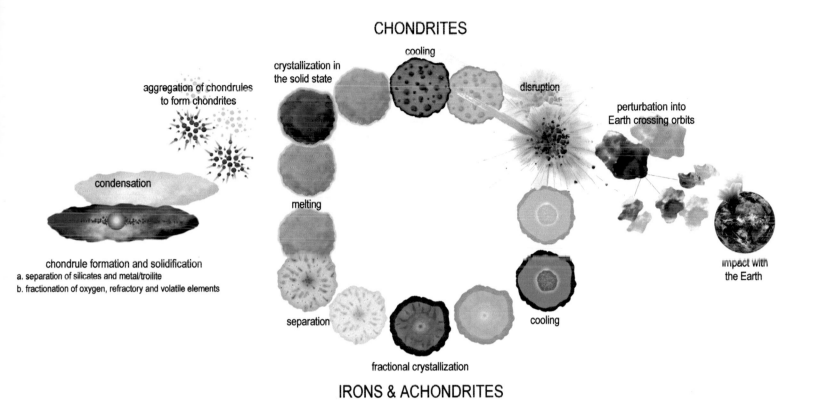

aggregation of chondrules
to form chondrites

crystallization in
the solid state

cooling

disruption

perturbation into
Earth crossing orbits

condensation

chondrule formation and solidification
a. separation of silicates and metal/troilite
b. fractionation of oxygen, refractory and volatile elements

melting

separation

fractional crystallization

cooling

impact with
the Earth

IRONS & ACHONDRITES

The evolution of chondrites and differentiated meteorites before fall to Earth.

meteorite, Canyon Diablo, yield a formation age that is identical to stony meteorites.

Other isotopic systems such as rhenium-187/osmium-187 and palladium-107/silver-107 used to date irons indicate that groups IAB-IIICD, IIAB, IIIAB and IVB have broadly similar ages, with a lower limit at around 4460 million years. Conflicting ages from different isotopic measurements in Group IVA irons have them alternately younger and older than other irons that crystallised from molten metal. Although it has yet to be resolved, the time of solidification of Group IVA irons differs from other groups.

Preservation of the relics of short-lived radioactivity in irons suggests that the separation of metal, and subsequent draining to form cores and solidification, all took place within a few million years of the assembly of their parent asteroids. An isotope of tungsten provides some supporting evidence of the timing of the separation of metal and silicate during core formation in asteroids. Decay of now extinct hafnium-182 with a half-life of 9 million years to tungsten-182 provides a link between high-temperature, oxygen-loving elements such as hafnium, and iron-loving elements such as tungsten. Tungsten isotopes are roughly the same proportion

in all magmatic irons, suggesting that metal drained from silicates over a short time of perhaps less than 5 million years.

Confirmation of the great antiquity of differentiated meteorites comes from the eucrite achondrite Piplia Kalan that fell in India in 1996. Recent measurements of excess magnesium-26 from the decay of extinct aluminium-26 in the calcium-aluminium rich feldspar of Piplia Kalan shows that the crust of the parent asteroid of the eucrites formed within about 5 million years of the birth of the Solar System.

Release into space

The third 'age' of meteorites is a measure of the time they were sprung from their parents and hurled unceremoniously into space. Travelling through the Solar System, small meteoroids are exposed to bombardment by cosmic rays. This extremely high-energy radiation emanates from both within and beyond the Solar System. Cosmic rays are essentially energetic protons with speeds approaching that of light. When they collide with solid rock their energy is such that they produce nuclear reactions. So intense are the impacts that they split the atoms they hit into new, lighter atoms.

Buried beneath more than a metre or so of rock on asteroids, the parent material of meteorites was shielded from high-energy cosmic rays. Break-up released many fragments into space, after which they gained a weak background radioactivity by interaction with cosmic rays. The amount of radioactivity depends on the length of time exposed.

Burial under about a metre of rock blocks even the most energetic cosmic rays, and the Earth's atmosphere is also sufficient to protect us from their potentially harmful effects. In space, however, prolonged exposure of fragments less than about 1 m across to cosmic rays produces a weak background radioactivity, and a range of secondary atoms. Even so, the number of nuclear splitting reactions is quite small, taking millions of years to generate only a few new atoms.

When a meteorite falls to Earth, shielding by the atmosphere shuts down the reaction. Rates at which some atoms are produced by proton bombardment are known from experiments. A direct measure of the 'exposure age' of a meteorite can therefore be obtained by counting the secondary atoms: the longer a meteoroid is exposed, the greater is the production of new atoms.

Measurements of cosmic ray-produced isotopes,

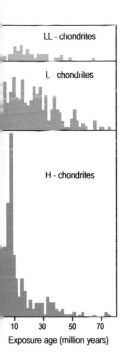

The exposure times of ordinary chondrite meteorite groups H, L and LL.

such as potassium-41 and potassium-40, reveal that some irons wandered the Solar System for up to a billion years before falling to Earth. Deep Springs, an iron with no known family, orbited for about half the age of the Solar System before landing on Earth in North Carolina. Ordinary chondrites have exposure ages varying between 1 million and 70 million years. Most irons have exposure ages ten times those of stony meteorites. Neon gas produced by cosmic rays in stony meteorites shows that many existed as small rocks in space for a few million, up to a few tens of millions of years. The weaker stones do not appear to survive more than about 50 million years in space before they are reduced to dust.

Exposure to cosmic rays not only gives an indication of the time meteorites drifted through space, but also reveals something of the manner in which their parent objects were broken up. For instance, the exposure ages of some iron groups cluster around a particular time. The asteroidal core represented by the IIIAB irons, for example, was largely destroyed about 650 million years ago. A similarly catastrophic, but more recent event released the IVA irons into space around 400 million years ago.

The L- and LL-chondrites have a wide range of exposure ages, between 10 and 50 million years, while a large number of H-chondrites were released around 8 million years ago. After the major collision 500 million years ago that expelled some of their argon gas, the further break-up of the L-chondrite asteroid was a more protracted affair as they were slowly chipped away by impacts. In contrast, the parent asteroid of the H-chondrites, or a large fragment of it, suffered a catastrophic collision releasing many fragments about 8 million years ago.

Down to Earth

When meteorites land on the Earth, their weak radioactivity — the by-product of their journey in space — starts the clock for their last 'age'. Now shielded by the atmosphere from cosmic rays, those

radioactive isotopes produced in space decay with their distinctive half-lives. The strength of this radioactivity gives an indication of the length of time a meteorite has been on Earth.

Isotopes used to measure the age of a meteorite's Earth 'residence' include carbon-14, with a half-life of 5700 years; and chlorine-36, with a half life of 300 000 years. By comparing the activity of suitable radioactive isotopes in a meteorite find with that in a freshly fallen meteorite of the same type, the time since it fell can be measured.

The activity of carbon-14 in some stony meteorites found on the Nullarbor Plain shows that they have lain on that ancient surface for up to 35 000 years. Other meteorites found at Roosevelt County, in New Mexico and the Sahara, fell up to 50 000 years ago. The lengths of Earth residence of Antarctic meteorites show the greatest variation: from a few tens of years to more than two million years, although most dated stony meteorites from Antarctica fell between 10 000 and 20 000 years ago.

Deeply altered remnants of a meteorite were found in limestone at Brunflo in Sweden in the early 1950s. The terrestrial age of this 'fossil' meteorite is the age of the rock in which it is entombed. Limestones at Brunflo formed during the Ordovician period, 460-480 million years ago. Identification of the meteorite in 1979 from resistant minerals surviving alteration show that it was an ordinary chondrite, among the most common types seen to fall. Brunflo's terrestrial age is many times greater than the average cosmic ray exposure age of ordinary chondrites falling today, and records a much earlier episode of break-up on their parent bodies.

In 1988, another fossil meteorite was discovered in cut slabs of limestone of Lower Ordovician age, around 5 million years older and some 500 km from the discovery at Brunflo. Further investigation of a 3-m section of the Holen Limestone in the Thorsburg quarry near Österplana in southern Sweden has so far given up 40 fossil meteorites. Fragments were found in an 80-cm thick layer, and may belong to the same meteorite fall. Another three pieces were found at different levels above this layer. The Österplana meteorites vary in size from about 1.5 to 8 cm across. Like the find at Brunflo, they have been almost completely altered to terrestrial minerals, but original chromite and a high iridium content confirm their meteoritic origin. A much younger fossil meteorite of Late Cretaceous age (around 65 million years old) was found in a sedimentary core from the north Pacific Ocean in 1996.

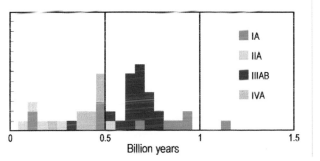

The exposure times of some of the iron meteorites from chemical groups IA, IIA, IIIAB and IVA.

FOUR AND A HALF BILLION YEARS IN A CHONDRITE'S LIFE

When deeply weathered fragments of the Coorara stony meteorite were picked from the Western Australian Nullarbor Plain in 1966, their transport to the Western Australian Museum in Perth was the last in a series of events spanning the age of the Solar System.

Coorara belongs to the low-iron ordinary chondrites, and like all meteorites of its kind, was assembled around 4555 million years ago. With a recrystallised texture, typical of meteorites commonly seen to fall today, the rock bears deep wounds suffered while resident on its parent asteroid. A microscopic view reveals dark veins snaking through an otherwise unremarkable chondritic texture. These sinuous black streaks, less than a millimetre thick, disregard all in their path. Made from melted and deeply disturbed minerals, the veins testify to high-pressure waves of energy that once swept instantaneously through the fabric of the rock.

Veins of such shock-altered rock are not unusual in stony meteorites, as they pervade many different types. In Coorara, just as in other meteorites, the cause of these dark invaders are past collisions in space. Although the veins appear small and insignificant, their contents tell volumes about conditions that prevailed at that instant in time.

In Coorara's dark veins there are a couple of unusual minerals. One has the same chemistry as the olivine from which it was made. Intense, but short-lived high pressure squeezed and compacted the atoms in the olivine to make a denser version, ringwoodite. First found in the Tenham meteorite that fell in Queensland in 1879, ringwoodite is a record of the enormous pressure experienced when the veins formed: nearly a million times Earth's atmospheric pressure. The second mineral, majorite, is a high-pressure equivalent of the other common meteoritic mineral, pyroxene, but with the atomic structure of garnet.

Coorara is not alone. Many other members of its family suffered similar impacts, some experiencing even more extreme damage. Once pale grey, these heavily battered rocks are now generally blackened by abundant melting. While the extent of the damage in the L-chondrites varies, they have one thing in common. During compressive heating caused by these impacts, they lost a large portion of their trapped gases, and the collisions can be dated to between 500 and 300 million years ago.

However, it was not until much later that the L-chondrites were freed as small, metre-sized fragments into space. As recorded by their exposure to cosmic rays, release of L-chondrite meteoroids in much the same sizes and shapes as they fall as meteorites today took place more recently, beginning around 70 million years ago. Gradual destruction of pieces of the L-chondrite asteroid, eventually throwing meteorites like Coorara on an Earth-bound course, was a more gentle event leaving no recognisable scars.

Measurement of radioactive carbon, produced by cosmic rays, in Coorara shows that it lay on the Nullarbor's surface for around 33 000 years after its fall. Now slightly rusted, the onslaught of weathering had begun to mask the meteorite's origin from space.

Carbon-14 dating of the Nurina 004 L6 ordinary chondrite shows that it lay on the surface of the Nullarbor for around 33 400 years before being found in 1986.

Below: The 8-cm long fossil meteorite (right) found in limestone in the Thorsberg quarry in Österplana, southern Sweden, is a chondrite. It fell into a warm shallow sea approximately 480 million years ago, and now little remains of its original minerals, which have reacted with the surrounding rock to produce a halo. To the left is a fossil nautiloid.

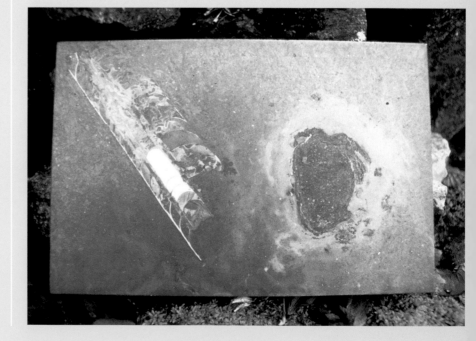

The Coorara L6 ordinary chondrite (right) has a marked black shock vein caused by a catastrophic impact in space. Under the microscope (above: field of view 4 mm), the vein is seen to consist of material blackened and melted by shock, and grains of ringwoodite (purple) and majorite produced by the transformation of the minerals olivine and pyroxene at high pressures.

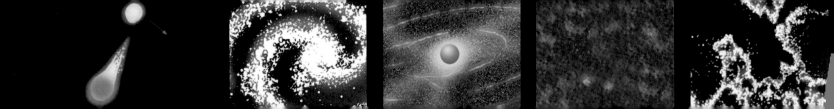

When you have excluded the impossible,

whatever remains,

however improbable,

must be the truth.

SIR ARTHUR CONAN DOYLE

DECODING THE MESSAGES

Conan Doyle's detective was a stickler for facts. Seen through the eyes of his companion Dr Watson, Sherlock Holmes' piercing scientific logic entertained readers of the *Strand Magazine* for more than 35 years. Quoted above is perhaps Holmes' most famous detective principle.

Scientists employ similar methods to decipher the mysteries of the natural world, although they cannot always be so incisive as Sherlock Holmes. Separating what we know to be true from that about which we can only speculate is the most important role of science. On Earth, at least, we can visit the 'scene of the crime' to gain more information. But meteoriticists are particularly hampered, for although, like their latter-day forensic colleagues, they can pick clues from the 'body' to analyse in the laboratory, in many cases, even if they could reach it, the 'scene' no longer exists.

We know that most meteorites are the debris from asteroids, but we do not know for certain which ones are represented in meteorite collections. Remote observations of asteroids continue to improve our knowledge of the make-up of the Main Belt and Near Earth Asteroids. Family relationships between some meteorites and possible parent asteroids are continually being suggested.

Recent close-up images of asteroids, such as Gaspra, Ida, Kleopatra, Toutatis and Eros, show them to be highly irregular objects with deep impact scars: testifying to long collisional histories. Ida even has a small satellite, called Dactyl, in orbit around it. Over the history of the Solar System, disturbances in the Main Belt and mutual collisions between asteroids have provided fragments, some of which have fallen to Earth as meteorites.

An important question to ask is: how representative of early Solar System materials are meteorites falling to Earth? When a census of meteorite families and orphans is taken, there may be pieces of more than 130 different asteroids in museum collections. Three quarters of meteorites seen to fall are ordinary chondrites, and perhaps represent debris originally from only three of these parent asteroids. Born of ordinary chondrite parents, only the group IIE irons appear partly to fulfil the family credentials.

Before the discovery of Antarctic and desert meteorite riches, the assumption that ordinary chondrites were the most abundant materials from asteroids appeared reasonable. A closer look at the asteroids themselves reveals a different picture. Asteroids showing all the outward features of ordinary chondrites are rare. On the contrary, of the possible parent objects of meteorites viewed spectroscopically, carbonaceous materials abound. The closer the asteroid belt is scrutinised, the more carbonaceous material is identified.

Similarly, amongst chondritic meteorites, most families reveal a carbonaceous allegiance. Of rare igneous meteorites, including ureilites, angrites and brachinites, more than half may have had carbonaceous chondrite parents. Abundant dust, thought to represent a much wider sample of the Solar System, contains only a tiny fraction resembling ordinary chondrites. Impact mixing of stony meteorites also shows a predominance of carbonaceous fragments. And ordinary chondrites disturbed by impacts with other bodies frequently contain foreign carbonaceous fragments, but only one case (Bencubbin) of the opposite is known.

The fact is that ordinary chondrites appear to be very rare in the Main Belt of asteroids. Why they

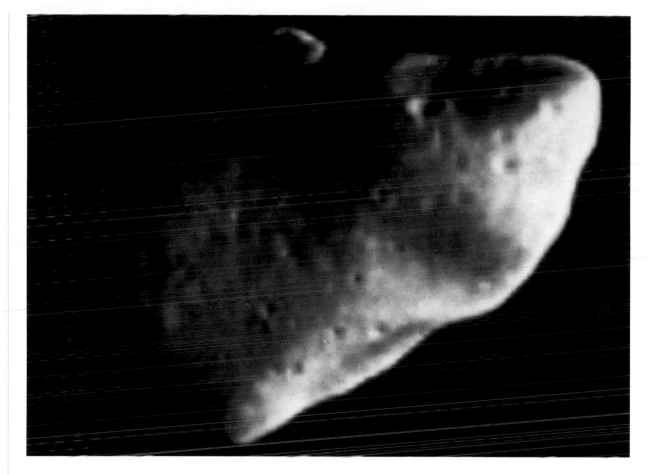

should be so abundant amongst meteorites is unknown. 'Population' statistics comparing ancient desert meteorites with modern falls suggest that the numbers and types of meteorites falling have not changed significantly for at least 50 000 years, although there is a possibility that some ordinary chondrites found in Antarctica came from different parent asteroids to those of modern falls. Fossil meteorites from Sweden show that ordinary chondrite-like meteorites have been falling to Earth for considerably longer than any recovered from deserts.

Various theories have been put forward to explain the disappearing act of the ordinary chondrites' parents. Perhaps ordinary chondrite asteroids have become disguised by space weathering, or else they are small and undetected. Measurements by the

Ordinary chondrites are the most common seen to fall today. Their asteroidal sources, however, remain enigmatic. Carbonaceous material appears to be the most abundant in the Solar System, but destruction of these fragile rocks may bias the meteoritic sample we receive on Earth.

NEAR-Shoemaker space probe on Eros provide support for the theory that it is made of ordinary chondrite material altered by space weathering. Even if they are rare in the Main Belt of asteroids, regular falls of ordinary chondrites show that they are certainly abundant amongst the objects crossing Earth's orbit. However, their importance to the history of the Solar System lent by their abundance on Earth may be overstated.

Presented with meteorites and dust, scientists attempt to solve the jigsaw puzzle of information they contain, trying to reconstruct the events that led to the birth of the Sun and the planets that surround it. Only meteorites record the crucial events that took place between 4570–4450 million years ago during the genesis of the Solar System. To complicate matters, because of the random way in which this material finds its way to Earth, the destructive effects of passage through our atmosphere, the rigours of the Earth's environment and the human factors involved in the recovery of meteorites, the sample is already fragmentary and biased.

Although matter rich in carbon may be the most abundant amongst the source materials of meteorites, many carbonaceous meteorites are weak, crumbly materials and their destruction in the atmosphere or on Earth may account for the small numbers of this type of meteorite recovered. Short of visiting the

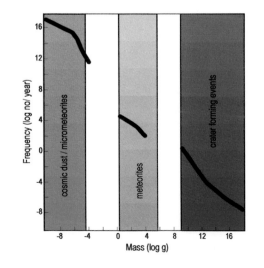

Left: The numbers of objects falling to Earth, ranging from tiny dust particles to asteroids with the potential to cause explosive impact craters. Calculations of the frequency of small meteorite falls indicate that it has not changed significantly in the last 50 000 years.

asteroids and comets, we have no direct means of knowing the exact way in which the different types of materials we see as meteorites are related or, with few exceptions, precisely the bodies from which they came. So we have to rely on some indirect sources of information.

Casting light on the subject

While meteoritic dust settles constantly on Earth, large, fragile objects with the same make-up could not survive passage through the atmosphere. These tiny particles are samples of material normally filtered by the Earth's atmosphere. Their sources are of intense interest and while their origin remains unknown, there are a few clues.

The intensity of the solar wind decreases with distance from the Sun, so the amount of damage that particles have suffered offers a clue to their source. At the Earth's distance from the Sun, the strength of the solar wind is about a hundred times greater than at Saturn. Minute crystal damage in some interplanetary dust particles is consistent with their origin from comets in the region of the Kuiper Belt.

Studies of sunlight bouncing off the surfaces of asteroids, compared with light reflected from meteorites and minerals in the laboratory, reveal some close matches. Around 14 kinds of asteroids are made of different materials and their arrangement in the asteroid belt is not random. Generally, 'bright' asteroids in the inner part of the Main Belt are similar to the igneous meteorites. These could be the sources of iron meteorites, pallasites and the basaltic achondrites. One asteroid, 4 Vesta, is a close match for the howardites and eucrites. Together with 20 small Vesta-like asteroids that may be the debris from one or more catastrophic collisions, Vesta may be the parent of howardites, eucrites and diogenites.

Darker asteroids dominate the central part of the asteroid belt. Possibly chondritic in composition, the surfaces of these asteroids appear altered by water. The outermost asteroids are dark, icy bodies apparently lacking water-bearing minerals. Although they may be rich in carbon, there is no known meteorite family to match them, but they may be the source of some chondritic interplanetary dust particles.

The fiery rain

Modern work shows that the majority of meteorites could only have come from a large number of small asteroids with independent histories. Jupiter's enormous gravitational influence over the asteroids essentially prevented accumulation of this early Solar System flotsam into a single body. Moreover, in the early Solar System, the region between Mars and Jupiter which is now relatively empty may have originally contained more material.

While the overall model of likening achondrites to crustal rocks, and irons and stony-irons to the interiors of small differentiated planets holds true, surviving chondrites do not fit easily into this scheme. Their mineral make-up and textures show that, since they formed, although some have been heated, they have not been melted and could not have undergone prolonged deep burial inside hot planets.

The Sun-like chemistry of chondrites, particularly CIs, and their extreme ages show that they were some of the earliest materials to form in the Solar System. It is probable then that they are some of the original materials from which the rocky planets and asteroids in the Solar System were built. It follows that chondrites can tell us most about the earliest events which led to the formation of the Solar System. Moreover, the origin of the chondrules themselves, and the high-temperature inclusions

Below: The formation of chondrules, such as those of the Saratov chondrite, and high-temperature inclusions holds many of the secrets to the origin of the Solar System.

with which they are associated, holds the key to our understanding of these early events.

Here we enter the realms of speculation and theory, for we do not know exactly from what, and by which process, those first droplets of Solar System material were produced. Nevertheless, studies of chondrites allow the construction of some plausible theories as to what the chain of events might have been.

For the formation of chondrules, within the last quarter of a century there have been at least 14 well argued theories, and a greater number of variations on similar themes. Even today, if you asked 14 meteoriticists about the origin of chondrules, although there would be strong similarities in their replies, in some respect you would get 14 different answers. While there has been no shortage of explanations, few seem up to the job. Gradually, many theories have been shown to be seriously flawed. Out of many theories two themes emerge: theories that postulate that chondrules formed by condensation directly or indirectly from a primordial cloud of gas and dust, the 'solar nebula'; or those that make chondrules by the melting of pre-existing solids in a variety of ways.

Common to all theories is that most chondrules formed from rapidly cooled liquid, or semi-liquid droplets. The key to the origin of chondrules is accounting for how a large number of melted droplets were made in a wide range of chemical environments in the Solar System over a very short time during its birth. In this, as our only tangible record of the events leading to the formation of solid material in the Solar System, the chondrites assume their true importance.

Perhaps, as the nebula gas condensed, droplets of liquid rock (the chondrules), and dust squeezed out of the cloud, cooled rapidly and began to accumulate into larger objects. This 'snowballing' effect continued until the bulk of the material was swept up, first into small planetesimals and then into planets. Alternatively, chondrules formed by collisions between particles in space, or melting of dust balls either by lightning discharges or shock waves in the primitive solar cloud. Still other theories suggest that chondrules formed during the collisions of protoplanets in the infant Solar System and are, in that sense, not as primitive as first thought.

All of these theories for the origin of chondrules have their problems, but after volcanism, now universally discredited, condensation is the hardest to defend. At low gas pressures in the early solar nebula,

the first materials to condense would have been mineral grains, not liquid droplets. Making chondrules from pre-existing dust is supported by observations. Some chondrules contain unmelted remnants of the material from which they were made. Dust is now accepted by many scientists as the raw material from which chondrules were manufactured, and remaining arguments focus on how the dust was melted.

Two of the greatest debates in meteoritics concern the sources of heat: firstly to produce the chondrules, and secondly to recrystallise or melt the parent asteroids of meteorites. Initially high temperatures are not only needed to form chondrules, but they are also required to drive the fractionation processes that produced chemical variation amongst chondrules and the overall compositions of the chondrite families. It is possible that in the different groups of chondrites we see, preserved at an early stage, the result of forces paramount in determining the nature of, and some of the differences between, the rocky planets of the inner Solar System.

Some suggested mechanisms for the origin of chondrules. No single theory has yet accounted for the production over a short time of a large number of liquid droplets in widely separated areas of the primitive solar nebula.

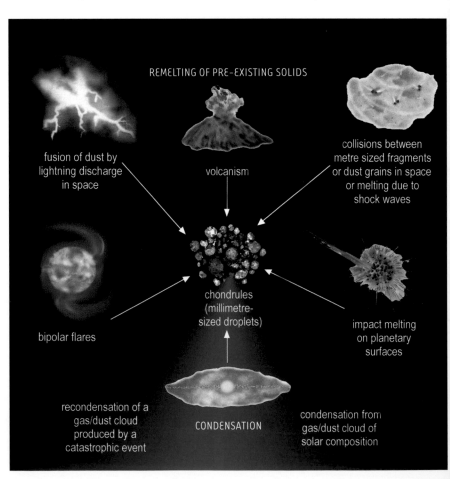

REMELTING OF PRE-EXISTING SOLIDS

fusion of dust by lightning discharge in space

volcanism

collisions between metre sized fragments or dust grains in space or melting due to shock waves

bipolar flares

chondrules (millimetre-sized droplets)

impact melting on planetary surfaces

recondensation of a gas/dust cloud produced by a catastrophic event

CONDENSATION

condensation from gas/dust cloud of solar composition

Meteorites give us information about events that took place during the earliest history of the Solar System. Steps in the formation of our Sun and its system of planets are thought to have started with the collapse of a molecular cloud in a small part of our galaxy, the Milky Way. In the resulting swirling cloud of gas and dust, the solar nebula, material started to condense. Dust grains clumped together, and were melted to form chondrules. Material accreted rapidly into larger bodies, eventually snowballing to become planetesimals and planets.

Three main mechanisms have been suggested to heat the parent asteroids of meteorites: the rapid decay of short-lived radionucleides such as aluminium 26; accretional energy from impacts; and electromagnetic induction. Of the three, radioactive heating and electromagnetic induction would have been the most efficient, and these mechanisms are generally favoured by meteoriticists to have produced the differentiated parent bodies of the achondrites.

From meteorites' ages of formation, we know that the major events in the genesis of the Solar System — the birth of the ancestral gas cloud, formation of high-temperature inclusions, formation of chondrules, the growth of asteroids and planets, and their subsequent melting — all happened within a relatively short time interval about 4570–4550 million years ago. And we can go back further than that. Although meteorites appear to have originated within the Solar System, their cargo of pre-solar

diamonds, silicon carbide and other tiny mineral grains originated from beyond and at various times before the birth of the Solar System.

From studies of young stars, astronomers observe that star pairs (binary stars) or multiple star systems are more common than previously thought. Now apparently isolated in space, our Sun may once have been part of a group of stars. Interactions between these juvenile stars and the massive clouds of interstellar gas and dust that surrounded them may account for some of the peculiarities of the Solar System. Evidence from pre-solar diamonds and the relics of now extinct radioactive elements in meteorites reveals that more than one supernova explosion or red giant star seeded the gas and dust cloud that eventually collapsed to form the Sun and planets. The most recent supernova, only a few million years before the formation of the planets, may have even triggered the event.

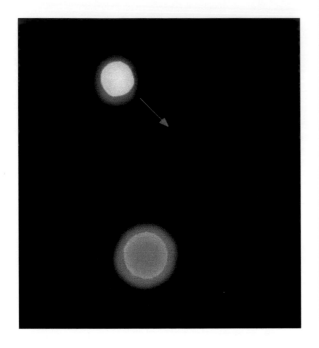

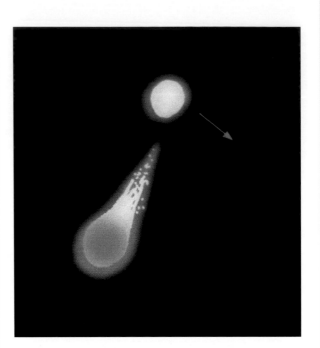

Left: Theories of a catastrophic origin of the Solar System are no longer supported. One such theory postulates another star passing close to the Sun and tearing off a filament of gas, which later condenses to form planets. While this could explain the distribution of momentum throughout the Solar System, calculations suggest that the filament would expand and disperse rather than condense.

Solar System beginnings

We do not know whether solar systems with a central star and orbiting planets are the exception or the rule in our galaxy, or the Universe as a whole. Numerous stars apparently have orbiting bodies that may be planets, but nothing remotely resembling our Solar System has yet been observed elsewhere in the galaxy. With nothing to compare it with, the Solar System remains one of a kind and theories of its origin are many and varied.

Of course, explanations for the origin of the Solar System must account for all the observed features. Two main approaches have been adopted to explain the origin of the Solar System. Firstly, there is the suggestion that planetary systems form as a natural consequence of the evolution of certain stars. Alternatively, our planetary system formed by a catastrophic event, such as the close approach of two stars, or the break-up of large proto-planets.

Since it was championed in 1796 by the Marquis de Laplace, the 'nébuleuse solaire' theory — that the Solar System formed from a primitive rotating cloud — natural evolutionary theories deriving the Sun and planets from such a cloud of gas and dust have generally been favoured. Transformed by crushing gravitational forces into a fiery thermonuclear inferno heralding the birth of a star, our Sun, material later condensed from the primitive cloud coming together to make the planets.

As the gas cooled, the cloud collapsed inwards and solid material condensed to form a disk around the Sun. Rotating more rapidly, the disk swirled until the outward centrifugal forces balanced the enormous gravitational attraction of the infant Sun. In stable orbit, the fragmental material in the disk was gradually swept up by collisions into larger bodies, ultimately forming planets.

A major problem with this otherwise elegant theory is that it predicts that most of the rotational energy (its angular momentum) should be centred in the Sun, which is opposite to that which exists. To account for this discrepancy, some scientists have favoured catastrophic theories for the formation of the Solar System.

One catastrophic theory envisages the close approach of two stars, a young diffuse star, our Sun, and a cooler, dense star. During the brief encounter, gravitational distortions between the two stars tore a filament of gas from the Sun. Because of the rotational energy gained from the passing star, the ejected material settled in orbit instead of being swept up again, and in much the same way as in the nebula theory, condensed to form the planets. Other catastrophic theories suggest more complicated, multi-stage collisional origins for the planets by the break-up of pre-existing giant proto-planets and recycling of the material.

Today, Laplacian theories for the origin of the Solar System are still favoured, although considerably modified from the original concept to account for the 'angular momentum' problem. Catastrophic theories are generally unpopular.

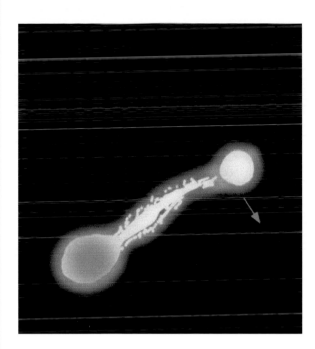 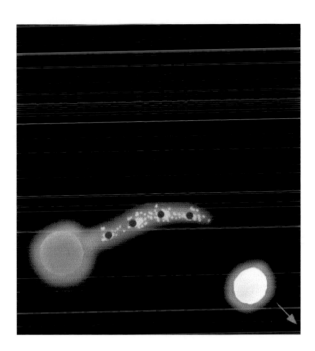

The view today

So how do meteorites fit into these various theories? Information gleaned from meteorites can be interpreted in many ways. The formation of the asteroids and planets by stages of condensation and accretion from an initially hot solar nebula, where the planets represent the last stage of the agglomeration of smaller bodies, is supported by the chondritic meteorites. They themselves appear to be the leftovers from the process. Unfortunately, no Earth-like chondrites exist and it is difficult to make the

inner planets from any meteoritic recipe. What follows is just one sequence of numerous possible events that may have led to the formation of the Solar System.

Depending on the accepted time of the Big Bang, the Solar System came into existence some 7-10 thousand million years after the formation of the Universe. By that time, quite late in its history, the Universe would have taken on an appearance not unlike the one we view today. In the outer spiral arm of one of a myriad of galaxies, the detachment of a small, insignificant patch of interstellar

Exactly how the chondrules formed remains unknown. One theory suggests that balls of dust were compressed and flash-melted by supersonic density waves set up in the unstable solar nebula and rotating through it faster than the orbiting gas and dust. In some parts of the nebula the waves compressed the gas, and heated the dust to melting point.

molecular cloud marked the beginning of what would be a remarkable series of events.

Several possibilities for the first step are equally plausible. But we know, from pre-solar grains in meteorites, that the molecular cloud was showered with diamonds from other stars. Perhaps because of its smallness and lack of rotational energy, the infant solar cloud was able to collapse quickly towards its centre rather than dividing, like an embryonic cell, into a binary star pair. This was an important step in the evolution of the Solar System, because whatever the actual series of events, it helped, ultimately, to determine its strange distribution of mass and angular momentum.

Overall, the pre-solar cloud contained 70 per cent hydrogen and 27 per cent helium, the small balance made of heavier elements. As the cloud collapsed towards the protostar, momentum gradually spread to the outer rotating disk of gas and dust. Several stages are suggested for the development of disks around protostars. An active stage marks the time when material falls rapidly towards the embryonic star and its captive disk. In the solar nebula, these energetic events led to the melting, evaporation and condensation of early formed, high-temperature solids. Like fly-ash, the white rock fragments of high-temperature minerals we now see in some chondrites cascaded through the cloud. This is the first event for which there is an accurate date: around 4566 million years ago.

In the regions nearest the forming star, early condensates settled towards the mid-plane of the disk. After the white inclusions, metal would have been the next to condense. Now made of gas and dusty grains of silicates, metal, sulfides and carbonaceous matter, considerable chemical and isotopic variations existed throughout the poorly mixed disk. Because of their magnetic properties and malleability, metallic dust particles may have clumped together more readily than silicates, their greater density carrying them to the mid-plane more quickly. In this way the apparent separation of metal from silicate, manifested in the different iron and oxygen contents of the chondritic groups, may have been effected.

During the early stages, virtually all the construction materials that could have built the inner planets were devoured by the protostar, leaving behind only a tiny fraction to complete the job. When the amassed material reached about a third of the present Sun, nuclear reactions were triggered. Shining brightly for the first time, the Sun made its appearance. With a rapidly rising surface temperature, the Sun grew to its present size and the forming system of planets felt the full blast of the solar wind. This outpour of energy stemmed the infall of gas, thus limiting the size of the star. As the collapse abated, the disc continued to orbit the juvenile star, which reached a hyperactive, super-luminous stage. Around this time, by whatever means, the chondrules formed. Flash-melting that produced the chondrules transformed, with ruthless efficiency, a large portion of the material we now see in chondrites.

One particular mechanism suggested for chondrule production provides such an energetic setting with abundant opportunities for melting, vaporisation and chemical fractionation over a short time. Instability in the solar nebula resulted in 'density waves' that rotated at uniform speeds compared with the rotating gas and dust. The density waves in the swirling disk may have resembled the spokes of a wheel emanating from the Sun. Close to the Sun, wave fronts swept the Solar System more slowly than the solid material, but at three times the Earth's distance from the Sun, the density waves rapidly passed through the more slowly moving dust and gas. Moving at around six times the speed of sound, these density waves compressed, heated and slowed the gas. Heated by friction with the gas, millimetre-sized balls of accumulated dust melted, the liquid droplets chilling rapidly to form chondrules in the rarefied wake of the shock waves.

Made predominantly of silicates, only rarely do chondrules contain much metal, but there are examples of metallic chondrules. Perhaps the early departure of much of the metal and sulfide to the mid-plane largely spared them the chondrule-making process. Early formed high-temperature grains apparently did not escape the hammer of the chondrule production line, and occasionally they were also melted to form chondrules, their chemistry pointing to earlier events.

Solid material eventually gravitated towards the mid-plane of the nebula where it agglomerated to form the complex rocks surviving today as chondrites. Exactly how chondrules and dust gathered first into fragments, then boulders, asteroids and eventually into much larger bodies is still very poorly understood — but aggregate they did. With remarkable rapidity, these events signalled the beginning of the major construction period of the Solar System. Spanning the time between 4555 and 4550 million years ago, these events are the second that can be accurately dated.

Temperature played an important role in the formation of the planets. In the hotter inner Solar System, the planets built their masses from rock and metal. At Jupiter and beyond, abundant ices were gathered along with rock. When Jupiter and Saturn reached critical mass they attracted huge gas mantles.

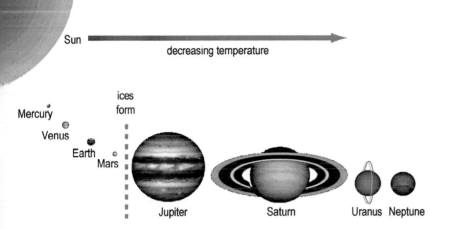

Sun ⟶ decreasing temperature

ices form

Mercury
Venus
Earth
Mars

Jupiter   Saturn   Uranus Neptune

Governed by the boiling points of the elements and their compounds, solid materials formed from condensation with slowly falling temperature in the solar nebula. The first minerals to condense are calcium- and aluminium-rich oxides, and silicates such as corundum and melilite. Later, grains of nickel-iron metal form, followed in quick succession by silicates such as olivine and pyroxene rich in magnesium, and calcium-rich feldspar. Volatile materials condense at much lower temperatures, when iron reacts with sulfur and oxygen to produce first troilite then magnetite. Finally carbonaceous materials and ices form.

Blown away by the vigorously active Sun, the inner part of the nebula was quickly purged of hydrogen and helium gas. Temperature-sensitive elements like lead, argon and xenon not already bound into rock, joined the exodus, leaving the embryonic bodies closest to the Sun deprived of these volatile components.

Around five times the current distance of the Earth from the Sun — far enough away from the star's heat for ice to accumulate — a colossal body was forming from the exported gas. Already ten times the mass of the Earth, the youthful Jupiter was dictating events in its part of the Solar System. With a gravitational pull great enough to trap the dispersing gas, this giant of rock and ice feasted on everything within its gravitational field.

Small planetesimals in the region of the asteroid belt were the first to feel Jupiter's grip. Many of those managing to avoid marriage with the bloated giant were banished from its kingdom, some to roam the outer reaches of the Solar System and beyond. Jupiter's tyrannical reign starved the diminutive Mars of a full complement of accreting material and prevented the construction of any substantial bodies in its immediate vicinity. The largest planet owes its bulk to a diet of asteroids. Today, Jupiter's legacy to us is a supply of primitive meteoritic rocks from the scrap yard of the asteroid belt, enabling scientists to decipher at least part of the complex history of the Solar System.

On the 'sunny side' of the asteroid belt, the parental bodies of meteorites were severely heated. Nearest the Sun, asteroids melted and separated into metallic cores and rocky crusts. At greater distances, gentle heating melted asteroidal ice with the result-

ing liquid water corroding and altering their constituent minerals.

Beyond the 'snow line' far out in the Solar System, an ice giant, Saturn, was growing. In the thinning nebula, much of which was dammed upstream in Jupiter, the planetary core of Saturn was unable to grow quickly enough to trap much of the rapidly-escaping gas. By the time Neptune and Uranus had built sufficient bulk, the gas was largely gone and they gathered mainly ices.

Back in the inner Solar System, scorched, rocky, embryonic planets jostled for position, gradually assembling, through catastrophic encounters, into four hardy survivors. The rough interplay left scars that we see today. Mercury may have been hit by a large body that stripped away much of its rocky coating. Another planetesimal probably clipped the proto-Venus sending it into a slow, reverse spin. Earth encountered a body about a third its size, the collision producing the Moon and tilting the planet on its axis in the process. The giant outer planets did not escape the mayhem. Uranus was knocked sideways such that it now rides around the Sun on its rings.

By 4000 million years ago, with the familiar features of the Solar System established, the inner planets suffered a cannonade of late accreting bodies that re-introduced some volatile elements, the pock-marked surfaces of Mercury and the Moon surviving to tell the tale. The stage was set for the appearance of life in the Solar System. On the water-rich planet, the third from the Sun, something stirred.

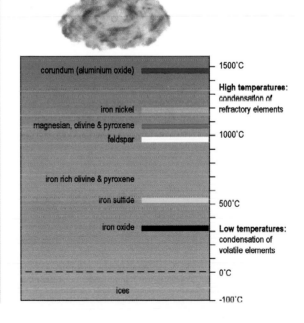

corundum (aluminium oxide) — 1500°C

High temperatures: condensation of refractory elements

iron nickel — 1000°C
magnesian, olivine & pyroxene
feldspar

iron rich olivine & pyroxene

iron sulfide — 500°C

iron oxide

Low temperatures: condensation of volatile elements

— 0°C

ices
— -100°C

Just as man, under the regard of the palaeontologist,

merges anatomically into the mass of mammals who preceded him,

so the cell, in the same descending series,

both qualitatively and quantitatively merges into the world

of chemical structure and visibly converges towards the molecule.

PIERRE TEILHARD DE CHARDIN

LIFE'S RICH TAPESTRY

Where, when, and how life began are questions to which there are, as yet, no answers. But evolve we did so another question is: from what? Two very different, but converging, approaches are used to track the possible origins of life. One is to go back in time and look in the most ancient sedimentary rocks on Earth for evidence of the earliest organisms. Another is to analyse pristine carbonaceous meteorites for clues of the original ingredients from which life may have arisen. Add astronomical observations of galaxies, laboratory experiments and theoretical molecular biology, and these are the only available lines of evidence.

3460-million-year-old fossil micro-organisms found in sedimentary rocks in the Pilbara of Western Australia are the oldest direct evidence of life on Earth. These simple-celled, but amazingly resilient organisms have persisted for more than three quarters of the Earth's history. More than 11 different kinds of filamentous microbes, some of which are structurally complex, have been described from these rocks — suggesting that replicating organisms had been around for far longer than the physical record would suggest. Today, in many quiet waters on Earth, active colonies of similar cyanobacteria trap sediment to build their domed homes called stromatolites.

Indirect evidence of life stretching back more than 3800 million years comes from the oldest known sedimentary rocks on Earth, found at Isua in Greenland and nearby Akilia Island. These ancient rocks contain chemical 'fossils' in carbon isotopes, suggesting the presence of life (if not its physical remains) when the sediment was laid down in water. This, then, along with fossil stromatolites, is the sum total of the available palaeontological evidence of the earliest life on Earth.

If we take the view that life evolved on Earth, the mystery of how it occurred remains buried in the dark years between the cooling of the Earth's crust around 4450 million years ago, and the first hint of the existence of organisms. The first 500 million years of the planet's history were turbulent times. Bombarded by huge asteroids and comets, the Earth's rocky molten surface and primitive atmosphere of carbon dioxide, methane, steam and ammonia probably supported no life. Intense bombardment by Solar System debris culminated in a cataclysm around 3850-4000 million years ago, the evidence of which is seen in the deeply cratered surface of the Moon.

Superheating by the Sun, combined with heavy cometary bombardment, may have stripped away the Earth's early 'cosmic' atmosphere. With time, however, an atmosphere and water were derived from the interior of the planet. As the Earth cooled, a crust formed and huge amounts of gas were exhaled from the solidifying surface. Steam condensed to form the first oceans. Additional water came from cometary bombardment. The early crust was probably made of basalt. Later, as the crust was repeatedly melted, lighter granitic liquid separated and rose to the surface, forming the precursors of continents.

Amidst this chaotic and seemingly inhospitable scene, life apparently arose. Since the fabrication of complex carbon compounds, including amino acids, in laboratory experiments during the early 1950s, the image of a warm pool of murky liquid rich in carbon, hydrogen and nitrogen, occasionally struck by lightning and bathed in intense ultraviolet light,

Fossil stromatolites from the Pilbara of Western Australia are among the oldest evidence of life on Earth. They were built by tiny micro-organisms similar to cyanobacteria around 3460 million years ago.

is inextricably linked with the possible origin of life. However, such conditions are not unique. If at least part of the process can be reproduced in the laboratory, and life did begin in an energised 'primaeval soup', there is a likelihood that it may have originated in more than one place, and at more than one time.

The fact is that the real nature of the earliest atmosphere is unknown. It is more likely that life began in a second-generation atmosphere — rich in carbon dioxide, water vapour and nitrogen — close to the end of the heavy bombardment of the Earth. In any event, the timing of the rise of life on Earth seems to hover around 4000 million years ago, with water as an essential ingredient.

So what role did meteorites play in all of this? Carbonaceous matter in meteorites was recognised 200 years ago. Since its discovery, great interest has focused on possible links between this cosmic material and the origin of life on Earth. Early work suggested that the substances in meteorites were formed by biological activity, pointing to the possibility of life in space. Modern investigations show beyond doubt that these early findings were either misinterpretations, or the result of Earthly contamination. Neither extra-terrestrial life, nor anything remotely like it, has yet been proven to exist in carbonaceous meteorites, although their carbon compounds are of extra-terrestrial origin. Because they originated beyond Earth, interest in meteoritic organic compounds remains high, as they could provide evidence of

crucial steps in the sequence of events that ultimately led to life.

Identifying the range of carbon compounds in meteorites, and establishing plausible ways for their manufacture in space, has pre-occupied many scientists. Unfortunately, the work is sometimes confounded by modern contamination. In the early 1960s amino acids were found in the Orgueil CI carbonaceous chondrite. These are characterised by the presence of amino (nitrogen and hydrogen) and carboxyl (carbon, oxygen and hydrogen) groups. Amino acids are the building blocks of proteins which are, in turn, essential ingredients of cellular organisms. Having lain in the collection of the Museum d'Histoire Naturelle in Paris for a century, the possibility of human contamination of Orgueil was high. Further analysis showed that the make-up of the amino acids in the meteorite was consistent with those from ten fingerprints. What, then, is the hard evidence?

On the one hand...

A breakthrough came with the fall of the Murchison CM2 carbonaceous chondrite in 1969 in Australia. Many fragments of the shower were quickly recovered, reducing the risk of contamination of their deep interiors. Finders were intrigued by the curious smell of methylated spirits emanating from the dark rocks. The smelly substances were later identified as aromatic hydrocarbons.

In what is now a classic laboratory experiment, Stanley Miller made amino acids in 1953 from an artificial atmosphere of ammonia, methane, hydrogen and water. He passed electrical sparks through the mixture and collected the products. The organic compounds produced contained a high proportion of amino acids, including a small percentage of the 22 amino acids that all creatures use to build proteins. However, the experiment is probably not a true representation of the chemistry of the pre-life Earth, although it shows what might have happened had these conditions existed.

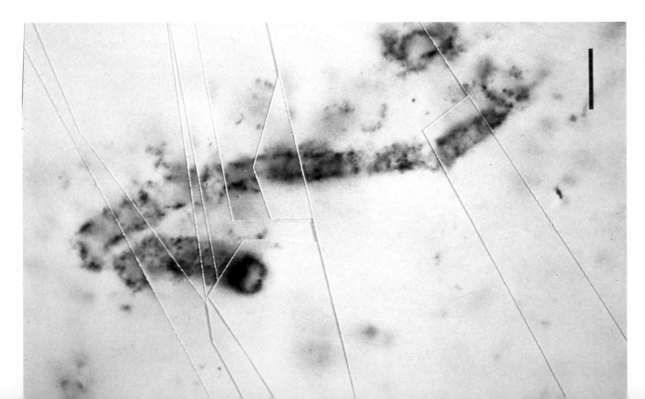

Fossil filament-shaped micro-organisms, also from the Pilbara, are the earliest direct evidence of life on Earth.

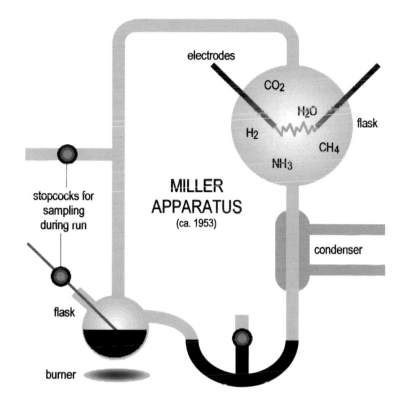

electrodes

CO$_2$

H$_2$O

flask

H$_2$

CH$_4$

NH$_3$

MILLER APPARATUS
(ca. 1953)

condenser

stopcocks for sampling during run

flask

burner

Deciphering the puzzle of how these vital molecules got their distinct twists is an important step towards understanding the origin of life.

Dominance of one spiral form of each biogenic molecule appears essential for life. If life began in one place, at one time, with both spiral forms of amino acids, perhaps each mutated differently until left-handed molecules out-competed the right-handed forms that disappeared. When and how the 'handedness' of molecules was selected lies at the heart of the mystery. If carbon-based life exists beyond the Solar System would it have the same handedness?

Exciting recent work on the Murchison carbonaceous chondrite, and another meteorite of the same type that fell at Murray in Texas in 1950, revealed an apparent excess of left-handed amino acids. This raises the possibility that left-handed amino acids dominated in the Solar System before the appearance of life on Earth, perhaps even before the formation of the Solar System.

Some important clues to a possible space-borne explanation for the handedness of molecules lies in the light streaming from regions of intense star birth in nebulae. In 1998, an infrared camera on the Anglo-Australian telescope trained on a region of the Orion Nebula detected circularly polarised infrared light, in which electromagnetic waves rotate steadily as they stream from young stars. Depending on the rotation direction, ultraviolet light polarised in this way could selectively destroy either right- or left-handed amino acids. If the infant Solar System was similarly bathed in twisty starlight, it could have broken the bonds in enough right-handed amino acids to leave a left-handed excess. Ultraviolet light was not detected in Orion, perhaps obscured by dust, so whether or not the die was cast for life's handedness in this way is still speculation. But it raises the intriguing possibility that in other molecular clouds in the Universe an opposite polarisation may have caused an excess of right-handed amino acids.

Amino acids were quickly identified in Murchison, but their structure revealed that these compounds were not formed by living organisms. Molecules of amino acids produced by organisms have the same structure: groups of atoms spiralling to the left around a central core of carbon. This 'left-handed' twist is always reproduced because it is dictated by the genetic code of cells which does not allow for variation. In contrast, the biogenic sugars in deoxyribonucleic acid (DNA) and ribonucleic acid (RNA) always spiral to the right. DNA is present in all chromosomes and is a component of the chemical basis of heredity. Amino acids from Murchison, and those manufactured in experiments, have nearly equal amounts of left-handed molecules and mirror image right-handed molecules, and this is the hall-mark of a non-biological origin.

The banded iron formations in sedimentary rock at Isua in Greenland were laid down more than 3800 million years ago, and testify to the presence of liquid water on the planet at that time. Carbon isotopes trapped within the rock indicate the presence of life.

One of the myriad questions yet to be answered is whether the left-handedness of amino acids is a pre-requisite for life, or an artifact of it. If life evolved from amino acids and other complex molecules brought to Earth by comets, meteorites and dust, then the selection of handedness may have preceded the first organisms. Many scientists find problems with this view, doubting that huge quantities of amino acids manufactured in space would have survived the journey to Earth without losing their dominant handedness.

Life springs eternal?

The discovery of indigenous amino acids and a rich cocktail of more than 400 complex carbon compounds of non-biological origin in meteorites has revolutionised views of how life on Earth might have evolved. While the complex materials found in meteorites such as Murchison cannot be interpreted as 'life', they represent some of the essential ingredients from which simple organisms might have arisen on the surface of the early Earth. Dubbed 'chemical evolution', the theory suggests that the precursors of life could have been made under primitive Earth conditions.

Other planets in the Solar System were also bombarded by cometary material shortly after they formed. Presumably, Murchison-like materials fell on their surfaces just as on Earth. Given the right conditions, could life have evolved independently from that on Earth? In August 1996, scientists from NASA announced that they had discovered tentative evidence of traces of 'fossilised bacteria' in the Allan Hills 84001 Martian meteorite from Antarctica, sparking unprecedented public and scientific interest in our supposedly 'dead' neighbouring planet.

If the findings are correct then the implications are enormous. Life could have evolved chemically on two planets in the Solar System. Alternatively, life might have been seeded from Mars to Earth, or the other way around. If the former were true, life could evolve easily on suitable planets anywhere in the Universe. The latter theory, sometimes called 'panspermia', suggests that primitive organisms spread like an incurable disease infecting anywhere life-containing meteorites land. However, earlier interpretations of the complex molecules in Murchison as 'life' have long since been discredited, consigning 'panspermia' rightly to the realms of science fiction.

No-one doubts the Antarctic meteorite's impeccable Martian credentials. But unlike other youthful Martian igneous rocks with ages of 180–1300 million years, ALH 84001 formed 4500 million years ago, shortly after the planet itself. Lengthy exposure to cosmic rays shows that the rock was hurled into space by an impact around 16 million years ago. In scientific circles, however, there is deep scepticism about the evidence of life within the rock. Given that the age of fall of the meteorite shows it lay in the Antarctic ice for around 13 000 years, there are strong suspicions that the 'life' evidence is, at best, Earthly contamination. Today the prevailing view is that all the observations in ALH 84001 are amenable to a non-biological origin.

Small, but perfectly formed

Rocks that solidify from molten magma are hardly the place to look for, or find, fossils. However, for most of its history ALH 84001 was part of the Martian crust where movements and impacts produced cracks along which fluids could have passed. Deeply eroded gullies and canyons sculpt the planet's surface, and are evidence of early water on Mars, with other features indicating perhaps even more recent liquid water activity. Even today a considerable amount of water (along with more abundant carbon dioxide) is locked in Martian polar ice.

The contentious evidence of the former presence of organisms in ALH 84001 lies bound in tiny globules of carbonate minerals, much less than a millimetre across. Scattered along cracks permeating the meteorite's fabric, these mineral beads may have grown from invading Martian ground-water rich in carbon dioxide. Each globule is rimmed by two concentric black bands containing sulfide minerals and random specks of tiny crystals of magnetite.

Magnetite (magnetic iron oxide) is a common mineral in meteorites and on Earth. Here it accounts

Right: The dark matrix of the Murray CM2 carbonaceous chondrite is rich in complex carbon compounds, including extra-terrestrial amino acids with a preponderance of left-handed forms, raising the possibility that they were abundant in the Solar System before the appearance of life on Earth. (Field of view 4 mm.)

Left- and right-handed amino acids.

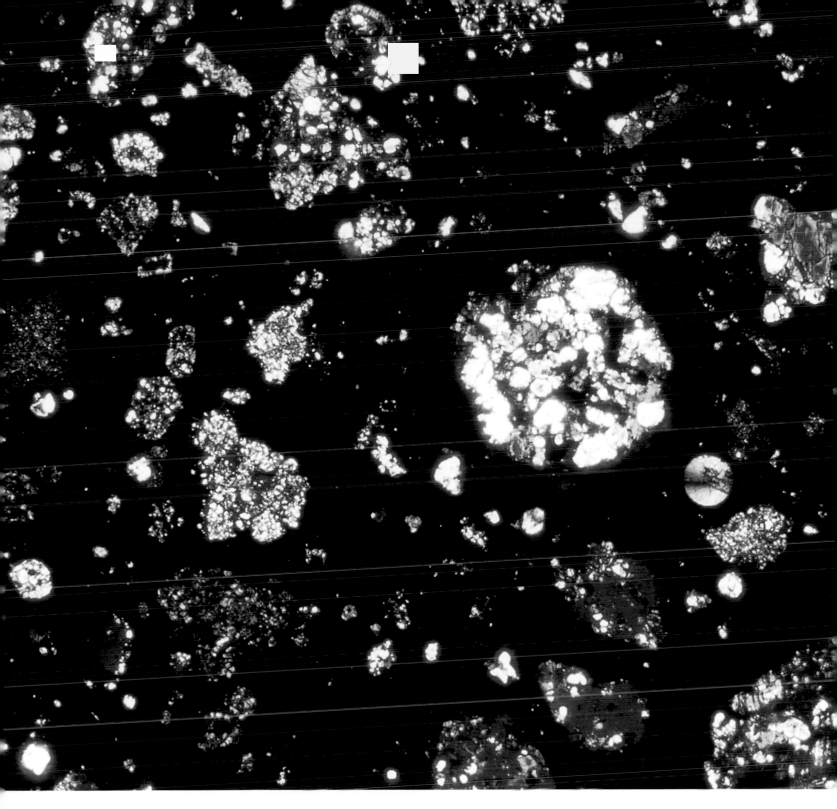

for a portion of the iron ore we manufacture into steel. But the magnetite crystals entrapped in carbonate globules in ALH 840001 are special for a number of reasons. They are less than one ten thousandth of a millimetre across, nearly chemically pure, and have flawless atomic structures. Such diminutive perfection is only known in magnetite crystals grown by unusual bacteria. In their quest for food these bacteria use their secreted magnetite and the Earth's magnetic field to sense which way is up and which way is down.

Not all of the crystals of magnetite in ALH 84001 hint at a biological origin. Many different shapes of magnetite are found in the meteorite, but so far only one kind cannot be explained by a non-biological origin. Tiny hexagonal columns of magnetite buried in the carbonate globules are identical to those made by bacteria and are presently unknown from any other process. But similarity is not proof of a causal relationship. Non-biological processes of which we have no notion may yet be responsible for the magnetite crystals.

The good oil

Complex carbon compounds abound in the carbonate globules in ALH 84001, although they themselves are not proof of the activity of organisms. One type, polycyclic aromatic hydrocarbons, are particularly abundant. Shortened to PAH (pronounced *pa*), these oily molecules are not produced or used by organisms. However, heating can convert dead organisms into PAH. The suggestion is that PAH in the meteorite may be the relics of baked Martian organisms.

PAH abound in other meteorites, such as carbonaceous chondrites, where they certainly formed inorganically. They are also abundant in cosmic dust, comets, and throughout interstellar space. PAH are everywhere on Earth, so could have contaminated ALH 84001 during its lengthy stay in Antarctica. Significantly, no other Antarctic meteorites or micrometeorites are as rich in PAH as ALH 84001, but then abundant carbonate globules have not been found in any other Martian meteorite. If the PAH are meteoritic hitch-hikers from Mars, they may not be evidence of life. PAH molecules in carbonaceous meteorites may have formed by reactions between gases in the pre-solar cloud. Reactions between carbon dioxide or carbon monoxide with hydrogen together with magnetite as a catalyst are known to produce PAH.

From carbon isotopic studies it turns out that most of the organic compounds in ALH 84001, and another Antarctic Martian meteorite Elephant Moraine 79001 (EETA 79001), are contamination from Earth. Amino acids extracted from ALH 84001 are typical of life on Earth. The remaining unexplained organic material does not exclude a biological origin, but does not necessarily support it.

The worm turns

Perhaps the most dramatic signs of the possible pre-existence of life in ALH 84001 are serried ranks of tiny worm-like structures. Originally interpreted as 'fossil bacteria', these curiously aligned bodies are like a cluster of organisms feeding on a carcass. But are they really fossil organisms? Because they are around one hundredth the size of bacteria on Earth, many scientists argue they are far too small to hold life's essential replicating ingredients. Additional arguments against the 'fossil' theory are that the structures are simply aligned crystals of magnetite, or the results of the preparation of samples for electron microscopy.

Ultimately, the size argument may not hold water. Tiny structures of similar size but different shape to those in ALH 84001, called 'nannobacteria' or 'nanobes', are found in ancient and modern sediments on Earth. Cocooned in an outer skin that may be a cell wall, nanobes recovered from rocks in Australia tested positive for DNA. However, their origin is hotly debated and the stains used to test for DNA are widely regarded as unreliable. Ultimately, nanobes may prove to be 'mineral', rather than 'animal' or 'vegetable'.

The proponents of life on Mars suggest that, taken all together, the evidence from ALH 84001 indicates the former activity of organisms. However, they freely admit that, individually, none of the observations is conclusive proof of their existence. Many other scientists claim that non-biological processes can explain the evidence. No-one disputes the observations, only the interpretations. Recent work has shown that many meteorites contain live microbes as a result of terrestrial contamination.

The lesson of the Martian controversy is that it is enormously difficult to prove the existence of life, let alone where it might have originated. For even if in the unlikely case that the 'worms' prove to be fossil bacteria, the spectre of Earthly contamination remains. Work continues on the curious structures and mineral associations in ALH 84001, with other Martian meteorites being similarly probed for additional evidence. So far the studies are inconclusive, but it has given impetus to NASA's plans to put astronauts on Mars early this century.

Below: The Martian igneous meteorite ALH 84001 found in Antarctica is said to contain evidence of life. (Black cube measures 1 cm.)

Deep gullies and ravines on the surface of Mars are abundant signs of once active water on that planet.

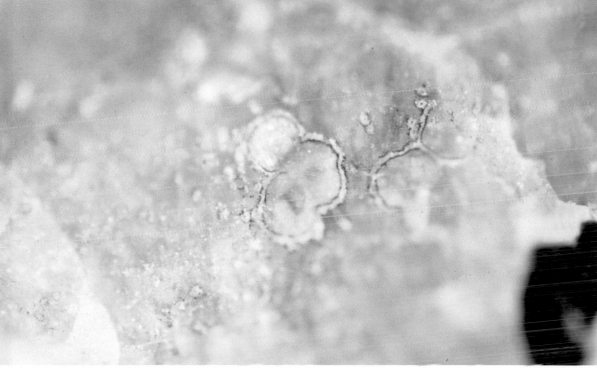

Carbonate globules (yellow) along cracks in the Martian meteorite ALH 84001, each only about 0.2 mm across, contain the contentious evidence of the possible existence of life on Mars. The dark rims contain magnetite and iron sulfides.

The Horsehead nebula
in Orion is a region rich
in dust, and the
spawning ground for
new stars. Complex
carbon compounds
rich in deuterium
relative to hydrogen,
similar to those found
in carbonaceous
meteorites, are
abundant in cool
interstellar molecular
clouds and inconsis-
tent with their
formation in the
Solar System. Some
of life's essential
ingredients might
have come from the
interstellar medium
before the Solar
System formed.

May they [the comets]

not one day come down entirely?

Shall we desire it?

They might sweep away cities and mountains

— deeply scar the earth and rear from their own

ruins colossal monuments of the great catastrophe.

BENJAMIN SILLIMAN, c 1840

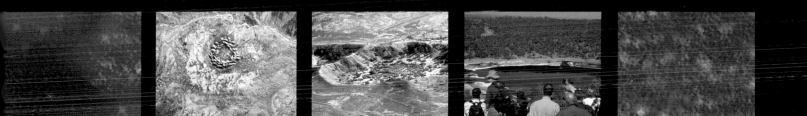

BLASTS FROM THE PAST

During the 1950s and 1960s the threat of nuclear war seemed very real. At the end of World War II the super powers pre-occupied themselves with maintaining a balance of weapons of mass destruction, ensuring they were never launched in anger. The alternative — unimaginable nuclear holocaust — was the deterrent. One of the great worries of our age ended with the removal of the 'iron curtain' and cessation of the arms race. Largely overlooked, however, a natural, indiscriminate threat to civilisation with potential destructive power much greater than any nuclear bombs still lurks in space — rogue asteroids and comets. Ironically, nuclear technology may one day help us to deal with them.

As far as the Earth is concerned, the furious period of accumulation of material that shaped the planet ended about 3800 million years ago. Since then, geological forces have taken over and the infall of extraterrestrial material has, by comparison with the mass of the Earth, dwindled to virtually nothing. But there is gathering evidence that massive stragglers in this mopping-up process occasionally punctuate our geological and biological history. Today, large chunks of this space debris still pose a threat to our existence. Astronomers estimate that there are more than 2000 asteroids bigger than 1 km in diameter near the orbit of the Earth. Less predictable, and potentially more destructive comets continually invade the inner Solar System. Trapped in this shooting gallery, the Earth has been hit many times in the past.

Around 200 sites where asteroids and comets have collided with the Earth are known on land and beneath the oceans. The scars of these encounters vary in size from a few tens of metres across to more than 100 km, and in age from a few thousand to more than 2000 million years. Most impact sites are now just ghostly circular outlines that no more than hint at their existence in ancient eroded rocks, whilst a few are clearly visible as large, deep craters.

These surviving impact craters provide the hard evidence against which predictions of the effects of potentially catastrophic impact can be compared. Estimates of the likelihood and frequency of impacts are gained from the impact record, together with an inventory of Earth-crossing objects and their orbits. How are craters formed? How often have catastrophic impacts occurred on Earth, and what effects have they had? Most importantly, can they happen again and what threat do they pose to civilisation?

Inside craters

Most of the time, the Earth's atmosphere is a formidable defence against space debris, but it is utterly defeated by large, tough asteroids. Atmospheric friction is not sufficient to slow bodies weighing hundreds or thousands of tons or more. These colossal projectiles retain a substantial proportion of their initial, cosmic velocity, hitting the Earth at speeds greater than 15 km/s. On impact, the release of immense stored energy can cause an explosion greater than the most powerful nuclear bomb.

Meteor Crater in Arizona, USA, is one of the world's best known and extensively studied meteorite impact craters. Formed about 50 000 years ago by the impact of a mass of meteoritic iron estimated to have weighed more than 50 000 tons, the bowl-shaped crater measures 1.2 km in diameter and 170 m deep. In 1927, an enterprising engineering company drilled exploratory shafts in the crater floor to locate and mine the large metallic 'asteroid' they believed to be buried beneath the crater. At that time, the way in which these craters form was poorly understood. Basic research revealed the futility of the venture.

When a massive meteorite travelling at high speed collides with the Earth, the body punches a hole, pulverising the rocks deep below the surface. In a fraction of a second, the projectile is stopped and the immense energy generated as a result of its enormous mass and speed is instantaneously converted to heat. Consequently, the projectile itself, and a portion of the surrounding country rock, is melted and vaporised. The resultant shock waves blast away the target rocks jetting debris upwards and outwards in every direction. Upturned strata, and a rim raised above the surrounding country, perfectly exhibited at Metcor Crater, are characteristic features of meteorite explosion craters.

Meteorite craters dot the surface of the Earth.

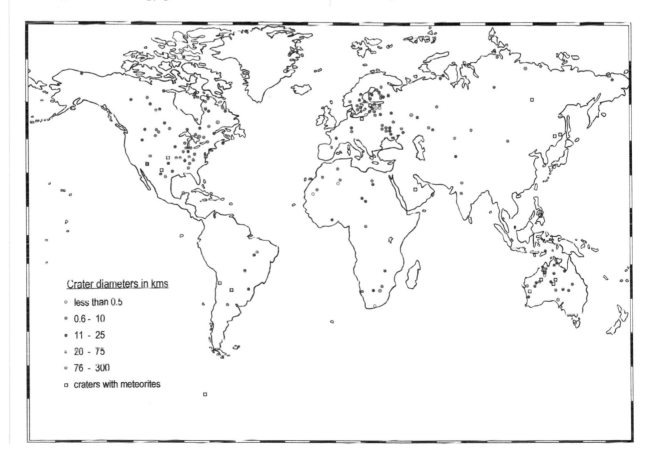

Crater diameters in kms
- ∘ less than 0.5
- ∙ 0.6 - 10
- • 11 - 25
- ▪ 20 - 75
- ● 76 - 300
- ▫ craters with meteorites

As the result of the formation of an explosion crater, the bulk of the meteorite is destroyed. This sets an upper limit (approximately 100 tons) to the size of meteorites expected to survive as single bodies, and for dynamic reasons fragments are rarely found within explosion craters. At Meteor Crater, more than 30 tons of meteoritic iron recovered from the crater rim and surrounding plain is testimony to its origin. Those meteorites found on the rim of the crater are twisted and deformed, shrapnel-like fragments.

After Meteor Crater, Wolfe Creek Crater in the north of Western Australia is the second largest impact crater with associated meteorites. More ancient and deeply weathered than Meteor Crater, Wolfe Creek Crater is partly overrun by sand dunes. Sophisticated dating techniques carried out on remnants of the impacting meteorite and the crater rocks show that Wolfe Creek Crater is around 300 000 years old. Other large bowl-shaped craters, such as Tswaing Crater (1.13 km across) near Pretoria in South Africa, Mount Darwin Crater (1 km across) in Tasmania, and Roter Kamm Crater (2.5 km across) in Namibia, are all of impact origin but lack surviving meteorites.

Wolfe Creek Crater and Meteor Crater are two of only 18 craters with associated meteorites known from around the world. Wolfe Creek is nearly circular, its diameter varying between 870 and 950 m. The outer slopes of the crater rise to form a rim that is at most 35 m above the surrounding sand plain. The inner walls plunge steeply, forming cliffs to the flat crater floor some 55 m below the rim. Because the bowl of the crater has become filled with wind-blown sand, today the crater floor lies only about 25 m below the level of the surrounding plain. Originally the crater was about 150 m deep.

During the formation of Wolfe Creek Crater, a large volume of rock disappeared, either pulverised and dispersed as a dust cloud, or vaporised. A little of this rock was hurled on to the rim and around the crater as single blocks weighing many tons. The variably deformed rocks in the crater walls suggest that the impacting projectile travelled from the north-east to south-west. Most of the meteorites at Wolfe Creek have been completely altered by weathering to iron oxides. Many 'iron shale-balls', some weighing tens of kilograms, are the deeply rusted remnants of the projectile. A few surviving pieces of fresh metal show that it was an iron meteorite.

Projectiles usually less than a few tens of metres in diameter produce 'simple', bowl-shaped craters

The Henbury craters in the Northern Territory of Australia were formed by the explosive impact of an iron meteorite about 5000 years ago. Australia has been populated for at least 40 000 years, and the formation of the craters may have been witnessed by Aborigines.

with raised rims, like Wolfe Creek and Meteor Crater. During the formation of a simple crater, the impact generates a shock wave and a release wave. These waves compress, melt, vaporise and excavate the country rocks creating a cavity. Then the cavity wall collapses inward, leaving a lens of brecciated rock surrounded by highly fractured, and sometimes melted country rocks.

At diameters usually above about 2 km in sedimentary rocks and 4 km in tougher crystalline rocks, impact structures take on a different form.

A splendid example is the 22 km diameter Gosses Bluff structure situated on the Missionary Plain 160 km west of Alice Springs in the Northern Territory of Australia. One of the obvious differences between Meteor Crater and Gosses Bluff, apart from size, is the presence of a centrally uplifted area in the latter. Structures like Gosses Bluff are called 'complex' structures and testify to truly catastrophic events in the geological past.

During the formation of complex structures, as with simple craters, the impacting projectile

Veevers crater (80 m diameter) was discovered in July 1975 in Australia.

excavates a cavity by melting and vaporising the country rocks. However, the recoil of the compressed rock beneath the point of impact lifts the cavity floor upwards. The rim of the cavity collapses to form the final crater which has a marked centrally uplifted area surrounded by a ring-shaped depression. This depression is filled with brecciated and pulverised rock, and pools of impact melt. The largest structures, like the 300-km diameter Vredefort Structure in South Africa, are even more complex. They are multi-ringed with circular ridges that spread from the central uplift like splash ripples in a pond.

The impact at Gosses Bluff occurred about 142 million years ago. Since that time the structure has been deeply eroded and only the deeper levels, well below the original crater, are preserved. Erosion has stripped away around 2-km depth of rock from the upper part of Gosses Bluff, and this presents a problem. In the absence of meteorite fragments as proof of their origin, what other than their circularity leads us to believe that these structures were formed by the impact of asteroids or comets?

The tell-tale evidence of a meteoritic origin for structures like Gosses Bluff falls into three main categories: structural, mineralogical and chemical.

Craters with associated meteorites

Name	Location	Diameter	Meteorite class
Eltanin	(submarine)	15 km	MES
Rio Cuarto?	Argentina	4.5 km	chondrite?
Meteor Crater	Arizona, USA	1.2 km	IAB
Wolfe Creek	Western Australia, Australia	880 m	IIIAB
Monturaqui	Chile	460 m	IAB
Macha	Russia	300 m	iron
Henbury	Northern Territory, Australia	180 m	IIIAB
Boxhole	Northern Territory, Australia	170 m	IIIAB
Odessa	Texas, USA	168 m	IAB
Kaarlijärv	Estonia	110 m	IAB
Morasko	Poland	100 m	IAB
Wabar	Saudi Arabia	97 m	IIIAB
Veevers	Western Australia, Australia	80 m	IIAB
Sobolev	Russia	53 m	iron
Campo del Cielo	Argentina	50 m	IAB
Sikhote Alin	Russia	27 m	IIAB
Dalgaranga	Western Australia, Australia	24 m	MES
Haviland	Kansas, USA	11 m	PAL

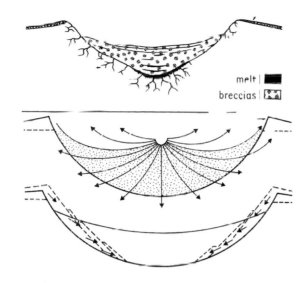

The formation of simple bowl-shaped craters, such as Meteor Crater, Wolfe Creek or Veevers.

melt |
breccias |

Geophysical surveys of many suspected impact structures show that, unlike volcanoes, they do not have deep-seated roots. At Gosses Bluff there is a limit to the depth of severely disrupted rocks at around 4 km below the present surface, indicating that the cause of the disruption could not have come from below.

Vital evidence that identifies impact craters from other similar geological formations is the presence of minerals altered or formed by intense pressure. The brief high pressures generated by the shock waves that excavate large craters cause transformations on microscopic scales to occur in certain minerals in the country rocks. The diagnostic features of true impact structures include: impact glasses from the melting of country rock; the rare minerals stishovite and coesite that formed by the compression of the mineral quartz; and microscopic planar features in quartz which in its unaltered form has no natural cleavage planes. Significantly, impact cratering is the only process known to produce these shock effects abundantly.

On a much larger scale, common features of intensely shocked rocks are shatter-cones. These are produced by the sliding of rocks along cone-shaped fractures. Like accusing fingers, the cones always point towards the impact. Finally, in many impact craters, gas and melt from the vaporised projectile was injected into the country rocks leaving a tell-tale chemical signature. Depending on the make-up of the meteoritic culprit, the nature of the signature can vary. In a few cases, by analysing the rocks in an impact structure for foreign material, it has been possible to match the signature with a known meteorite group.

The bowl of the Tswaing crater (formerly known as Pretoria Saltpan) in South Africa is just over 1 km across.

Although now deeply eroded, the Gosses Bluff impact structure reveals its true form when viewed from space. A central uplift made of resistant sandstone forms the bluff, but the ghostly outer rim, with a diameter of 22 km, can be seen on the surrounding plain.

The formation of complex impact structures,
such as Gosses Bluff.

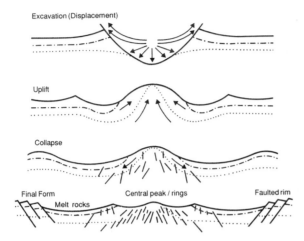

Excavation (Displacement)

Uplift

Collapse

Final Form   Central peak / rings   Faulted rim
    Melt rocks

In some impact structures, like Popigai in Russia
(100 km across) and the Nördlinger Ries structure in
Germany (24 km), the country rocks are peppered
with tiny diamonds formed by the compression of
graphite (carbon). Some may even have condensed
from vapour generated by the impact. Since they
survive over great periods of geological time, impact
diamonds are additional useful indicators in the identification of impact structures.

Above: A grain of quartz
from the Woodleigh
impact structure in
Australia shows planar
deformation features
(dark lines) that formed
as the result of high-
pressure shock waves.
(View about 2 mm.)

Left: Earth movements
subsequent to the
impact which formed
the Vredefort structure
in South Africa
disturbed the shatter-
cones from their
original positions.
When re-oriented,
the shatter-cones
point to the centre of
the structure.

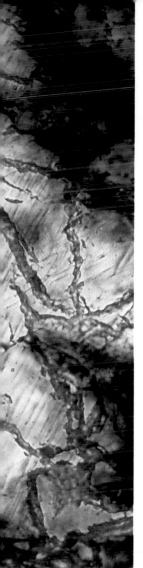

Evidence of giant impacts occasionally presents
itself in places other than at the site of the collision.
Rocks thrown great distances from craters landing in
alien terrain, and flying glass, such as tektites, remain
as incriminating evidence of impacts. Asteroids hitting seas and oceans raise giant tidal waves thousands
of metres high that break on land, depositing muddy
sediments and other debris as they flow and ebb
across hundreds of kilometres.

TEKTITES: SPAWN
OF GIANT IMPACTS

Tektites are not meteorites, but they have a fascinating history linked to catastrophic impacts in the geological past. These fragments of natural silica-rich glass are found in a few well-defined strewn fields in several parts of the world. Tektites formed by the melting of Earth's rocks at the point of impact between projectiles and target rocks. Highly compressed gases hurled them high into the atmosphere, and some landed great distances from the impact sites.

The first recorded tektites were found in the late eighteenth century in Czechoslovakia. Known as moldavites, they are around 15 million years old. Other tektites are found in southern China, Vietnam, Laos, Cambodia, Thailand (indochinites), Philippines (philippinites, or rizalites when from Rizal), Malaya (malaysianites), Indonesia (javanites from Java, billitonites from Billiton Island), Borneo, Australia (australites), Texas (bediasites) and Georgia (georgiaites), USA, and the Ivory Coast, West Africa.

Glass fragments that are 65 million years old from the Caribbean Island of Haiti are the oldest tektites known, and were launched by the massive Chicxulub impact on the Yucatan Peninsula of Mexico. The moldavites were splashed from the Nördlinger Ries impact event 15 million years ago, the Ivory Coast tektites are related to the 1.3-million-year-old Lake Bosumtwi Crater in Ghana, and the North American tektites originated from the Chesapeake Bay impact structure 34 million years ago.

Around 770 000 years ago, after an asteroid or comet impacted somewhere in Indochina, tens of thousands of tektites, some weighing hundreds of grams, fell over south-east Asia and Australia. Tiny beads of glass, microtektites, from the same event have been raised from the bed of the Indian Ocean from as far away as Madagascar. Collectively known as the Australasian tektite strewn field, this shower of glass is the largest and youngest known.

Additional melting of tektites during high-speed flight through the atmosphere produced symmetrical shapes. Many Australian tektites exhibit these shapes to perfection and sometimes have flanges formed by molten glass swept back from their direction of flight. The only other flanged tektites, other than australites, have been found in Java, and a single example dredged from the Indian Ocean.

The source of Australasian tektites is unknown, but is probably somewhere in eastern Indochina. A small number of Australian tektites come from a much older impact that occurred 10 million years ago, also at an unknown impact site.

An impact event in Indochina created the Australasian tektite and microtektite strewnfield.

This flanged button tektite was melted to glass, and sculpted by its high speed transit through the atmosphere when it was hurled more than 4000 km from its source, in a catastrophic impact in Indochina around 770 000 years ago.

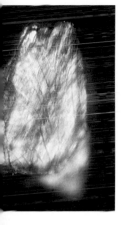

Above: Impact diamonds are known from only 12 craters worldwide, including the Nördlinger Ries crater in Germany. (About 0.5 mm across.)

Left: Blocks of country rock ejected from the impact that formed the Lake Acraman structure, South Australia, were found embedded in 600 million year old sedimentary rock 300 km to the east of the site of impact.

Above: In rocks from Clear Creek North, Colorado, the boundary between the Cretaceous and Tertiary periods is marked by a layer of clay (light band) covered by a thin 'fireball' layer (yellow) containing iridium and shocked grains of quartz, and capped with a band of dark coal.

Target Earth

Many new impact sites have been recognised in recent years, partly as a result of satellite photography and extensive geophysical exploration. Australia and other ancient continental landscapes, such as the Canadian Shield, Scandinavia, Siberia and Africa are particularly fertile hunting grounds. As our knowledge of impacts improves, a picture emerges of their awesome power and the role that these events might have played in the geological evolution of the Earth. There is also some evidence that asteroidal or cometary impact may have, at various times in the

past, played a dominant role in Earth's biological history. Locked away in ancient sediments on Earth, tantalising clues point to at least one impact that may have caused a global holocaust.

The challenge of this particular crime is that it happened around 65 million years ago. The victims, it is suggested, may have been a large number of creatures, including the dinosaurs. Fossil evidence for the cause of the extinction of around 70 per cent of the Earth's organisms at that time is disputed, but evidence for a catastrophic impact with global effects

is unequivocal. The first evidence discovered for this event lies in a thin layer of clay that was deposited at many places throughout the world 65 million years ago, at the end of the Cretaceous geological period.

Rarely more than a centimetre thick, the clay marking the boundary between the Cretaceous and the succeeding Tertiary geological period is unusually rich in the rare metal iridium: with between 30 and 300 times the average iridium content of the crust. Of the Earth's total budget of iridium, less than 0.001 g/ton of rock remains in the crust. Along with other metallic elements, most iridium sank into the Earth's metallic core very early in the planet's history. So where did the iridium in the clay layer come from?

Intense volcanic eruptions and giant meteorite impact are the most likely sources. Since meteorites contain more iridium than the Earth's crust, the impact hypothesis gained many early supporters. Also in strong support of an impact origin, the clay layer contains shock-altered minerals, carbon residues indicating global fires, micro-diamonds, droplets of tektite-like impact melt, and nickel-bearing oxides typical of those formed during the ablation of meteoritic bodies.

A corroded tektite fragment enclosed in a shell of iron-rich clay from deposits in the Caribbean island of Haiti.

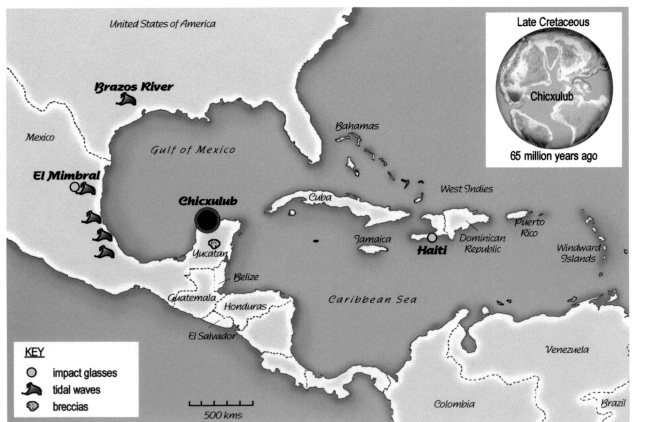

The Chicxulub impact structure on the Yucutan Peninsular of Mexico is now confirmed as the site of the catastrophic impact at the boundary of the Cretaceous and Tertiary periods. Related features of the event include deposits from tidal waves in the USA and Mexico, glass fragments in Haiti and possibly Cuba, and a boundary layer of the same age at many places around the world.

Deposits 3 m thick at Arroyo el Mimbral in Mexico were laid down by tidal waves generated by the impact at Chicxulub.

To account for the total amount of iridium in the clay layer world-wide (around 500 000 tons) it was suggested that an asteroid or comet 10–12 km in diameter collided with the Earth. The explosive power of such an impact would be greater than the simultaneous detonation of all the nuclear weapons currently held in the world's arsenals. If it struck the land surface, the body would punch a hole in the Earth's crust lifting more than 10 000 km³ of dust into the atmosphere. If the asteroid plunged into an ocean, the giant wall of displaced water would flood the coastlines of most continents. Fires triggered by the impact would send hot gases and soot, together with the dust, high into the atmosphere blocking the life-sustaining energy of the Sun.

At first, the Earth's surface would be plunged into freezing darkness, the food chain would be disrupted and large numbers of animals and plants would die.

Later, the insulating layer of dust in the polluted atmosphere would cause a 'greenhouse' effect with temperatures rising dramatically. Gradually over millions of years, as the dust settled, the atmosphere would cool, and the creatures and plants that survived could flourish once again.

Crucible of the catastrophe

Missing from the Cretaceous–Tertiary boundary was evidence of the site of the impact. Estimates of the size of the projectile suggested that the impact would leave a crater of the order of 150–200 km in diameter. At the time of the discovery of the 65 million year old metallic anomaly in the late 1970s, a couple of suitably sized candidates were known, but there was no crater precisely of that age. Those craters close to the right age were far too small.

A quest for the 'smoking gun' began. Scientists sought clues from the boundary clay. Logic dictated that the clay might be thicker closer to the point of impact. A search for obvious thickening of the clay was slow and painstaking. Eventually, a substantial thickening of the clay (2 cm instead of the normal 1 cm) was found in the USA. Along with the iridium-rich layer was a claystone layer with shock altered quartz grains that were a little larger than at other sites in the world. This pointed to an impact somewhere in the Americas.

Exposed along the Brazos River in Texas is an unusual, thin band of sandstone capped with a layer of clay coinciding with the Cretaceous-Tertiary boundary. Flow structures in the clay, and lumps of clay mixed with the sandstone are typical of catastrophic deposits caused by giant tidal waves. The nature of the disturbance indicates a wave about 100 m high. On the assumption that the original wave may have been 5 km high, scientists estimated that it must have been generated within about 5000 km of the Brazos exposures, perhaps in the Gulf of Mexico or the Caribbean.

Some of the most important clues to the location of the impact site were unearthed in Haiti. Here the boundary layer is 50 times thicker than at other sites, packed with spherules, tektite-like glass fragments and large grains of shocked quartz. The spherule layer is capped with a few millimetres of iridium-rich clay. The thickness of the layer suggested that the impact site was close, less than 1000 km away — but not from the present position of Haiti. In the late Cretaceous, the rocks that were eventually raised to form Haiti and the other islands of the Greater Antilles lay a thousand kilometres to the south-west of their present position, under 2 km of water. The search for the impact scar started from this previous location.

Corroborating evidence came from another site at Mimbral in north-east Mexico. Here a deposit from a tidal wave, with glass fragments at the base and an upper part with backwashed fossil charcoal, lay below a layer with diamonds, iridium and fragments of shocked quartz.

A buried circular structure, around 180 km in diameter, straddling the tip of the Yucatan Peninsula in Mexico, had been known since the 1950s. Thinking it a likely trap for oil, the structure was the target of Mexico's national oil company, which drilled a number of exploratory wells. By the 1960s, with no oil in sight, it was dismissed as volcanic. In 1978, two years before the first announcement of the iridium anomaly in the Cretaceous-Tertiary boundary clay, a re-evaluation by oil geologists had it marked as a possible impact structure.

Following the recognition of the Haitian impact glasses, scientists turned their attention to the Yucatan structure. Shock-altered quartz was soon found in drill cores, along with breccias and glass. Because they formed at the time of the impact, the glasses held the key to the age of the structure. Radiometric dating showed that they formed 64.98 ± 0.05 million years ago. Subsequent dating of the Haitian glasses at the Cretaceous-Tertiary boundary gave an age of 65.01 ± 0.08 million years — indistinguishable within experimental error from the the age of the crater. The 'smoking gun' had been found. Later the crater was christened after the small fishing port of Puerto Chicxulub, close to the centre of the impact.

To be, or not to be?

Incredible as it may seem, within an instant, the impact of the projectile at Chicxulub opened up a gaping hole in the Earth's crust around 80 km across and 30 km deep. Friction between the projectile and the country rock melted material that spewed from the back of the cavity. As the projectile was flattened, compressed, melted and vaporised, the cavity became lined with impact melt. Widening to its final diameter of 180 km, the cavity rapidly collapsed, becoming backfilled by giant landslides. Within seconds of the impact, like some colossal creature gasping for air, the rocky heart of the structure burst through the sea of melt, having rebounded through 20 km of crust. Dust, solidified droplets and vapour were sucked back up through the hole that the projectile ripped in the atmosphere. Spread out around the world, the finest particles of projectile and target eventually settled as the the iridium-rich layer of clay at the Cretaceous-Tertiary boundary.

The meteoritic case for the exotic metals and other materials at the end of the Cretaceous is now widely accepted, but what role the impact played in the extermination of the dinosaurs and other forms of life is hotly disputed by many highly sceptical palaeontologists. Although the dinosaurs dominated life on Earth for more than 150 million years, towards the end of the Cretaceous their numbers had dwindled to about a dozen isolated species. In evolutionary terms it would have taken very little to wipe them out completely. Curiously, other reptiles, like the crocodiles, lizards, frogs, turtles and many other land-living species of animals, not to mention the mammals, were little affected by whatever caused the disappearance of the dinosaurs.

From the fossil record, palaeontologists have shown that the birds are an early evolved, specialised branch of the dinosaurs that persist to the present day. But whatever the cause, the disappearance of the giant reptiles both on land and in the sea at the end of the Cretaceous opened the way for another group of creatures: the mammals. Fortunately for us, from that early stock, human beings evolved 63 million years later.

TERROR AT
TUNGUSKA

Right: Even in 1929, the forest levelled by the Tunguska event of 1908 still looked devastated.

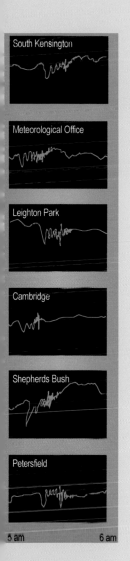

Between 5 and 6 am, on 30 June 1908, micro-barographs from various locations around England recorded the aerial shock waves caused by the atmospheric explosion over Tunguska.

Although an explosive, crater-forming impact on Earth has not been recorded in historical times, in 1908 a large chunk of Solar System debris posted a warning to the world of the potentially catastrophic effects of such impacts. On 30 June of that year, a huge explosion laid waste to a remote forested area around the Tunguska River in central Siberia. Eyewitnesses saw an object brighter than the Sun leaving a column of smoke in its wake as it plummeted through the sky, finally bursting into a plume of flame. The noise of the explosion was heard up to 1000 km away, and a man seated 60 km from the site was knocked flat by the searing blast.

It was not until nearly twenty years later that a Russian scientific expedition located the site. Even such a long time after the event, the scene of devastation was difficult to comprehend. Save for a group of decapitated, but upright, trees at the centre, an area of forest about 30 km across had been largely felled. Uprooted trees lay in a radial pattern pointing away from the centre of the blast and many were charred. All the evidence pointed to the explosive impact of a giant meteorite, but the investigators were puzzled as to why neither crater, nor remnant of the projectile could be found.

Many theories have been proposed to explain the Tunguska explosion, but few stand up to the evidence. In 1957, microscopic particles of the meteoritic culprit were found in soil taken from the site, indicating that the object must have exploded in the atmosphere. In laboratory explosive tests, Russian scientists were able to reproduce the pattern of devastation and concluded that the body exploded 5–10 km above the Earth. But what was the nature of the body?

The most plausible explanation is that it was a fragment of a comet that had rounded the Sun on a collision course to Earth. Masked by the Sun's glare, the comet would not have been observed by astronomers. To a massive, but crumbly, projectile like a comet, travelling more than 40 times faster than a high velocity bullet, our dense lower atmosphere would act like a solid wall, destroying the body before it hit the Earth. Another suggestion is that the body was a stony asteroid that suffered the same fate. In any event, calculations show that the down-blast at Tunguska was thousands of times the power of the atomic bomb dropped on Hiroshima in 1945.

Dust from the Tunguska explosion spread throughout the atmosphere producing unusually bright nights over Europe and Asia for two months after the fall. The temporary damage to our atmosphere caused by this airborne material included the

production of acid rain, depletion of the ozone layer and a small change in temperature. However, on an astronomical scale, the Tunguska explosion was insignificant, and the damage negligible.

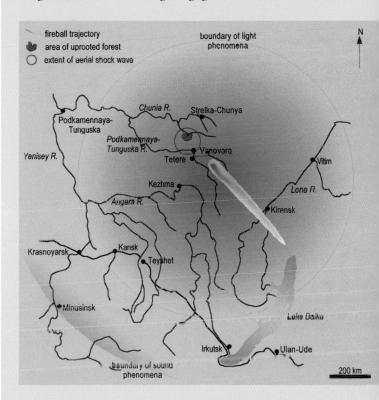

At the centre of the impact area near the Tunguska River in Siberia, the forest was devastated by the downblast of the enigmatic object that exploded in the atmosphere in 1908. The trajectory of the fireball passed close to the town of Vanavara, where an aerial shock wave was felt. Light and sound phenomena were seen and heard many hundreds of kilometres from the site.

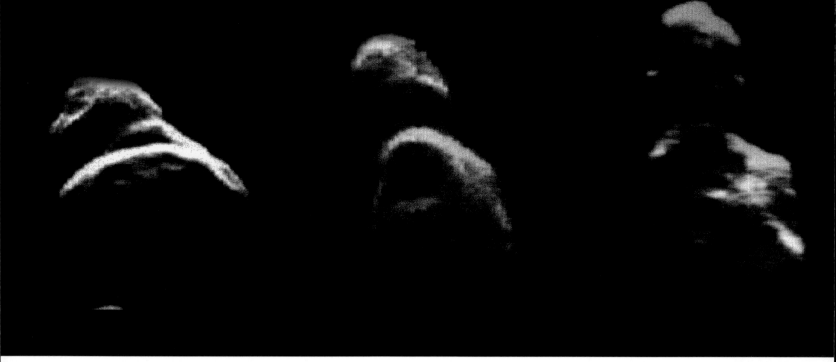

Can it happen again?

Studies of meteors and meteorites tell us that small meteorites, ranging from dust particles to objects much less than 100 g, are far more abundant than large ones. How often, then, do potentially destructive meteoroids strike the Earth?

Surprisingly, exploding fireballs (bolides) with the destructive power of a Hiroshima nuclear bomb occur once a year. Most of these go unnoticed because they occur high in the atmosphere, with little or no effect on the ground. Theoretically, a meteoroid weighing tens of thousands of tons, big enough to produce craters the size of Wolfe Creek and Meteor Crater, should arrive on Earth about once every 5000 years.

Ground impacts of small asteroids 250 m in diameter releasing more than 50 000 times the energy of an Hiroshima bomb in producing a crater several kilometres across, happen every 10 000 years or so. Devastating though they would be, their effects would be localised. Truly global catastrophic impacts occur on a very much longer time span.

Crater-forming impact on the scale of Meteor Crater has not occurred in historical times. The reason is that the Earth is not without its natural defences, and there are a number of factors that

work in our favour. Most meteorites are brittle objects and break up in the atmosphere to fall as showers of small objects rather than one large one. Craters with surviving meteorites show that most were made by the impact of iron meteorites that are far less common than stony meteorites and less prone to break-up in the atmosphere.

When all of the mitigating factors are taken into account, impacts producing craters on the scale of Wolfe Creek and Meteor Crater are predicted to occur perhaps once every 25 000 years, while collisions on the Popigai scale only about once every 15 million years or so. Potentially global catastrophic events, such as the Chicxulub impact, that might cause biological extinctions may only occur every 50-100 million years.

Realistically, what are the chances of catastrophic impact on Earth? If the fossil cratering record is anything to go by, over the vast time-scale of geological history, impacts with global effects have certainly occurred and are likely to happen in future. Humanity's written history is very short, a few thousand years or so, and this pales into insignificance against the 4500 million years of Earth's geological history. But because cratering has not occurred during historical times does not mean that it cannot happen again.

Asteroid 4179 Toutatis, an Apollo asteroid 6 km long, was discovered in 1989. Its orbit crosses that of the Earth: in 1992 it came within 4 million km; on 29 September 2004 it will pass within 1.5 million km. The chances of Earth colliding with a similar object during the next century are estimated at 1 in 30 000.

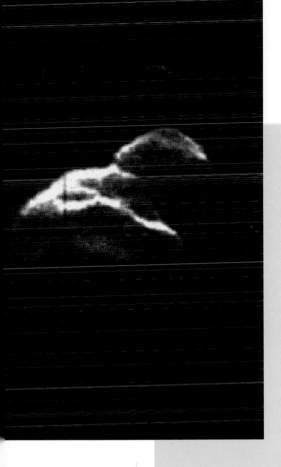

Top 20 meteorite craters of the world

Name	Location	Diameter (km)	Age (million years)
Vredefort	South Africa	300	2023
Sudbury	Ontario, Canada	200	1850
Chicxulub	Yucatan, Mexico	180	65
Manicouagan	Quebec, Canada	100	214
Popigai	Siberia, Russia	100	35
Woodleigh	Western Australia	>90?	?
Acraman	South Australia	85-90	580
Chesapeake Bay	Virginia, USA	85	35
Puchezh-Katunki	Russia	80	175
Morokweng	South Africa	70	145
Kara	Russia	65	57
Beaverhead	Montana, USA	60	>600
Tookoonooka	Queensland, Australia	55	128
Siljan Ring	Sweden	55	368
Charlevoix	Quebec, Canada	54	360
Kara-Kul	Tajikistan	45	>5
Montagnais	Nova Scotia, Canada	45	50
Araguainha	Brazil	40	247
Mjølnir	Norway	40	144
Saint Martin	Manitoba, Canada	40	220 ± 32

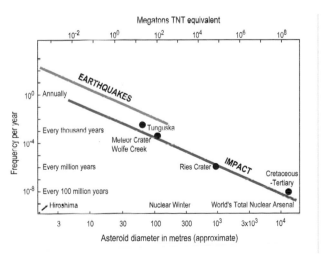

Frequency of meteoritic impact events, and comparable Earth events, with particular sizes related to energies (megatons).

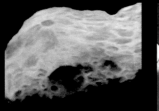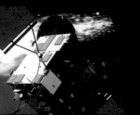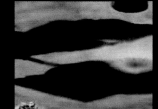

of the molecules of which we are composed.

This is perhaps the hardest truth,

for it allows no rest.

M CALVIN

TO THE FUTURE

Meteorites tell us a lot about the early Solar System, but there is still much to learn about the way in which the planets were constructed, and even more about the advent of life on Earth. Random and rare as they are, fresh meteorite falls are a boon to science, continuing to provide us with important information. Increasing interest in the space sciences has seen a surge in the number of new meteorites recovered, opening many new lines of research. Scouring the world's deserts for meteorites continues at a pace never seen before, while planned space missions will fill huge gaps in our knowledge.

So far only a small fraction of the world's deserts, including the immense Sahara, has been searched closely for meteorites. The Nullarbor is a prolific source of meteorite finds, but around 4 million km^2 of the Australian continent is equally dry. Central Australia may also contain abundant and as yet undiscovered meteorites. In fact the whole of the continent is probably harbouring one of the highest concentrations outside of Antarctica.

For every few hundred common meteorites found, an unusual, or previously unknown, type shows up. Studying these rarities will undoubtedly increase our knowledge of the Solar System. Even weathered finds of common meteorite types, heavily altered by the Earth's environment and seemingly unsuitable candidates for research, have the potential to tell us about changing climates in those desert regions where they have lain for millennia. Their concentration and terrestrial ages tell us something about the numbers of meteorites falling to Earth through time.

Despite the huge haul of meteorites from deserts over the last 30 years, the orbits of these ancient and modern falls cannot be determined — although their exposure to cosmic rays reveals something about their history in space. Plans are well advanced for an all-sky camera network designed for the harsh conditions of the Nullarbor in Western Australia, and this should be operational in 2003. Using a new generation of solar-powered, computer-driven cameras covering an area of 150 000 km^2 with low annual cloud cover, the expectation from known fall rates is to record at least three to four falls per year, each depositing meteorites bigger than a walnut.

In most terrains, even if it was known roughly where they were, searching for meteorites this size is extremely difficult. On the bare expanse of the Nullarbor, past recoveries show that spotting small meteorites is well within the capabilities of experienced searchers. A party of five requires around two days to cover an area of 1 km^2. Given the accuracy of photographically triangulated falls, around 4 km^2 would need to be searched thoroughly for each recorded fall, taking perhaps no more than a month to recover the annually expected crop.

Expanding this camera network to larger but similarly barren areas of the Australian outback, 20–25 falls per year, including two to three meteorites the size of grapefruits could be recorded. Operating over several years, recovery of only one in three of these meteorites pin-pointed by cameras vastly increases the number of falls with known orbits. The rarest of catches, a meteorite with the briefest exposure to cosmic rays could, by its immature orbit, betray the parent asteroid from whence it came.

Like sentinels, diligent astronomers continue to scan the skies for Earth-crossing and possibly civilisation-threatening asteroids and comets. So far, preparations to deal with this potentially catastrophic eventuality are woefully inadequate. Strategies to

The spacecraft STARDUST will gather samples of interstellar dust, and a cometary sample as it flies past Comet Wild-2 in January 2004, before returning to Earth in 2006.

combat a small Solar System body on a collision course with Earth rely on its destruction or, more realisitically, altering its orbit with nuclear weapons. The technology to do this with confidence of success is not yet available. Given that adequate warning of the possibility of such an event would strongly influence the outcome, we rely heavily on astronomical installations searching for potential Earth impactors. However, world-wide, there are currently few astronomers involved in this work.

As humanity stands on the threshold of this millennium, the prospect of another great era in the advancement of space science looms large. Taken out of context and far removed from their original sources, only so much can be learned from meteorites and dust particles. The next leap forward is to visit the asteroids and comets themselves, viewing their materials in the context they were made. Unmanned and remote exploration of small bodies in the Solar System has already begun in earnest. By early this century it is very likely that samples from an asteroid, a comet and a planet such as Mars will be returned to Earth for laboratory study.

In February 1999, NASA's STARDUST probe was launched on a 7-year mission taking it to a comet and back. Destined to fly past the nucleus of Comet Wild-2 in January 2004 at the relatively low speed of 6.1 km/s, STARDUST will collect particles from the comet on special panels. Using a substance called 'aerogel', panels coated with this unusual material are designed to trap tiny high-speed particles gently. Stuck in the remarkable aerogel, cushioned like a ball in a catcher's mitt, precious cometary particles will drop to Earth in a re-entry capsule in January 2006.

The Rosetta Stone, an inscription of an Egyptian proclamation written in three languages, allowed scholars to decode Egyptian hieroglyphs. The Rosetta Space Mission by the European Space Agency will fill huge gaps in our knowledge of comets. By sampling comet 46P/Wirtanen, the Rosetta space mission will study the origin of comets and what they can tell us of the origin of the Solar System.

In June 2003 the European Space Agency will launch the Mars Express mission. After a journey of six months, the small Beagle2 lander will be dropped from the main craft and head for the surface of the planet. Its mission, to answer the question: Is there, or was there, life on Mars?

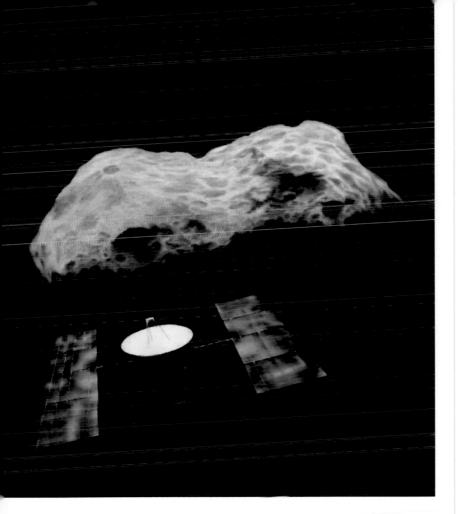

he MUSES-C mission, conducted by the Japanese Institute for Space and Astronautical Sciences with NASA assistance, s scheduled for launch n either December 2002 or May 2003, aiming to arrive at a near-Earth asteroid in September 2005. It will collect up to three surface samples before returning to Earth in June 2007.

During its three orbits of the Sun, STARDUST will also gather interstellar dust particles invading the Solar System, including a recently identified stream from the direction of Sagittarius. These particles range in size from a thousandth, to one ten thousandth of a millimetre and the prediction is that more than 100 grains will be collected.

Back on Earth, improved and highly sophisticated equipment is already under construction to analyse these tiny grains and determine their ages. Handling cometary dust at sub-zero temperatures, avoiding terrestrial contamination and other Earthly damage will challenge scientists and available technologies to their limits.

Another cometary mission, ROSETTA, is scheduled for launch by Europe's Space Agency in January 2003, intercepting comet 46P/Wirtanen in November 2011. The probe is named after the famous Rosetta Stone found in 1799 near Alexandria. Inscribed with a decree in three languages — hieroglyphs, demotic (late Egyptian) and Greek — the stone allowed scholars to decipher ancient Egyptian. On its 8-year mission to the comet, ROSETTA will pass close to two asteroids, Otawara and Siwa. Although ROSETTA will not return to Earth, it will use on-board equipment to reveal much about the composition and make-up of small Solar System bodies, helping to decipher their origin.

An asteroid sampling mission, Muses-C, is scheduled to be launched by Japan's Space Agency in November 2002. Aimed at the small near-Earth asteroid 1998 SF36, the purpose of the mission is to retrieve powdered or fragmental material from its surface. Encountering the asteroid in 2005, the orbiting space vehicle will release a probe, which will make two or three brief touchdowns on the asteroid, at each location firing a 5-g bullet into the surface. In this way an estimated 1 g of debris blasted from soil or bedrock will be collected. Re-united with the mother ship, the probe and its cargo of asteroidal fragments should return to Earth in June 2007.

Heightened interest in Mars, stimulated by Martian meteorites and the possible evidence of early life there, has made it the target for several future space missions. Mars Express, an ambitious unmanned European mission, is set to be launched after 2003. Part of the Mars Express' payload is Beagle2. Named in honour of the small ship which carried naturalist Charles Darwin in 1831–36 on the exploratory trip that inspired *The Origin of Species*, Beagle2 bristles with imaging, sensory and analytical equipment. Its primary aim is to search for evidence of past, or current, life on Mars.

Planned to target areas of Mars moulded by the past action of liquid water, which perhaps contains evidence of biological activity, the precise landing site of Beagle2 has yet to be chosen. Among the lander's many planned activities are the search for water and the presence of organic residues. Sniffing the Martian atmosphere, Beagle2 will seek traces of gases indicative of living organisms. Burrowing beneath the dusty Martian surface to a depth of 2 m, a robotic probe is designed to investigate the nature of Martian rocks, their chemistry, mineralogy and age.

Results from Beagle2, to be remotely transmitted to Earth, are eagerly awaited. Confirmation or denial of life, past or present, on Mars will shape the nature of future searches there and elsewhere in the Solar System. Other bodies marked as potential harbours of life include Europa, a moon of Jupiter. Although essentially rocky, Europa has a water-ice crust concealing an ocean of water below. But it will be some considerable time before detailed measurement can be carried out on this distant body.

Peering deeply into space, the Hubble Space Telescope will provide us with a view of our cosmological past. When all of this future work is taken into account, perhaps then we may have a better understanding of the central mystery of the origin of life. It seems very unlikely that the chance events that gave rise to the major features of the Solar System, let alone the rise of life on Earth, could be repeated exactly elsewhere. One day we may have to face the possibility that, in the universe, we are utterly alone.

The Hubble Space Telescope, named after astronomer Edwin Hubble, is a joint European Space Agency and NASA project. At 600 km above the Earth's surface it does not suffer from interference from the atmosphere. Hubble has already made some of the most important discoveries in the history of space science.

Small meteorites should impact on Mars at survivable speeds. Based on its low weathering rates, calculations suggest that the Martian surface could be littered with as many as 500 marble-sized meteorites per square kilometre. Whether or not a futuristic Mars colony is ever established, missions to the surface could recover much meteoritic material from a different sample population to that now available to scientists on Earth.

Comet Kohoutek,
photographed in 1974

SOURCES, FURTHER READING

Ancient beliefs

Alpher, B. (1969). The Egyptian hieroglyphic group. *Meteoritics*, 4, p. 132.

Bell, L. (1969). The Egyptian hieroglyphic group. *Meteoritics*, 4, pp. 131-32.

Bevan, A. W. R. and Bindon, P. (1996). Australian Aborigines and meteorites. *Records of the Western Australian Museum*, 18, pp. 93-101.

Bjorkman, J. K. (1973). Meteors and meteorites in the ancient Near East. *Meteoritics*, 8, pp. 91-130.

Buchwald, V. F. (1975). *Handbook of Iron Meteorites*. University of California Press, Berkeley, California, 1418 pp.

Burke, J. G. (1986). *Cosmic Debris: Meteorites in History*. University of California Press, Berkeley, California, 445 pp.

Dietz, R. S. and McHone, J. (1974). Kaaba stone: Not a meteorite, probably an agate. *Meteoritics*, 9, pp. 173-75.

Gettens, R. J., Clarke, R. S. Jr and Chase, W. T. (1971). Two early Chinese bronze weapons with meteoric iron blades. Smithsonian Institution Occasional Papers, 4, no. 1, 77 pp.

Ghafur, A. Sheikh (1953). From America to Mecca on airborne pilgrimage. *The National Geographic Magazine*, 104, pp. 1-60.

Graham, A. L., Bevan, A. W. R. and Hutchison, R. (1985). *Catalogue of Meteorites*, 4th edition. British Museum Natural History and University of Arizona Press. 460 pp.

Kahn, M. A. R. (1938). On the meteoritic origin of the Black Stone of the Ka'bah. *Popular Astronomy*, 46, pp. 403-407.

Krinov, E. L. (1960). *Principles of Meteoritics*. Pergamon Press, London-New York, 535 pp.

La Paz, L. (1969). Hunting meteorites: Their recovery, use and abuse from Paleolithic to present. *University of New Mexico Publications*, No. 6, University of New Mexico Press.

Marvin, U. B. (1996). Ernst Florens Friedrich Chladni (1756-1827) and the origins of modern meteorite research. *Meteoritics and Planetary Science*, 31, pp. 545-88.

Petrie, W. M. Flinders, Wainwright, G. A. and Mackay, E. (1912). *The labyrinth Gerzeh and Mazghuneh*. School of Archaeology in Egypt. University College, London, 60 pp.

Pillinger, C. T. and Pillinger, J. M. (1996). The Wold Cottage meteorite: Not just any ordinary chondrite. *Meteoritics and Planetary Science*, 31, pp. 589-605.

Roth, A. M. (1993). Fingers, stars and the 'opening of the mouth'. *Journal of Egyptian Antiquities*, 79, pp. 57-80.

Ross, J. (1819). *A Voyage of Discovery ... Exploring Baffin's Bay and ... a North-West Passage*, John Murray, London.

Heide, F. and Wlotzka, F. (1995). *Meteorites Messengers from Space*. Springer-Verlag, Berlin, Heidelberg, New York, 231 pp.

Sears, D. W. (1978). *The Nature and Origin of Meteorites*. Monographs on astronomical subjects: 5. Adam Hilger Ltd, Bristol. 187 pp.

Stanley, G. H. (1931). On the meteorite at M'bosi, Tanganyika. *South African Journal of Science*, 28, pp. 88-91.

Thomsen, E. (1980). New light on the origin of the Holy Black Stone of the Ka'ba. *Meteoritics*, 15, pp. 87-91.

Wainwright, G. A. (1932). The coming of iron. *Antiquity*, 10, pp. 5-24.

Wainwright, G. A. (1912). Pre-dynastic iron beads in Egypt. *Revue Archéologique*, 19, pp. 255-59.

Wainwright, G. A. (1932). Iron in Egypt. *Journal of Egyptian Archaeology*, 18, pp. 3-15.

Waldbaum, J. C. (1980). The First Archaeological Appearance of Iron and the Transition to the Iron Age. In Wertime, T. A. and Muhly, J. D. (eds) *The Coming of the Age of Iron*. New Haven and London, Yale University Press, p 69.

Wasson, J. T. and Sedwick, S. P. (1969). Possible sources of meteoritic material from Hopewell Indian burial mounds. *Nature*, 222, pp. 22-24.

Wayman, M. L. (1987). On the early use of iron in the Arctic, Part 1; meteoritic iron. *CIM Bulletin*, 80, No. 899, pp. 147-50.

Wynn, J. C. and Shoemaker, E. M. (1998). The Day the Sands Caught Fire. *Scientific American*, November.

Rocks from space

Bevan A. W. R., McNamara K. J. and Barton J. C. (1988). The Binningup H5 chondrite: A new fall from Western Australia. *Meteoritics*, 23, pp. 29-33.

Bland, P. A., Smith, T. B., Jull, A. J. T., Berry, F. J., Bevan, A. W. R., Cloudt, S. and Pillinger, C. T. (1996). The flux of meteorites to the Earth over the last 50 000 years. *Monthly Notices of the Royal Astronomical Society*, 283, pp. 551-65.

Brownlee, D. E. (1994). The origin and role of dust in the early solar system. In Zolensky, M. Wilson, T. L., Rietmeijer, F. J. M. and Flynn, G. J. (eds). *Analysis of Interplanetary Dust*, AIP Conference Proceedings 310. American Institute of Physics, New York. pp. 127-43.

Halliday, I., Blackwell, A. T. and Griffin, A. A. (1989). The flux of meteorites on the Earth's surface. *Meteoritics*, 24, pp. 173-78.

Halliday, I., Blackwell, A. T. and Griffin, A. A. (1991). The frequency of meteorite falls: Comments on two conflicting solutions to the problem. *Meteoritics*, 26, pp. 243-49.

Halliday, I., Blackwell, A. T. and Griffin, A. A. (1996). Detailed data for 259 fireballs from the Canadian camera network and inferences concerning the influx of large meteoroids. *Meteoritics and Planetary Science*, 31, pp. 185-217.

Hughes, D. W. 1981. Meteorite falls and finds: some statistics. *Meteoritics*, 16, pp. 269-81.

Keay, C. S. L. (1992). Electrophonic sounds from large meteor fireballs. *Meteoritics*, 27, pp. 144-48.

Shearer, C. K., Papike, J. J. and Rietmeijer, F. J. M. (1998). The planetary sample suite and environments of origin. In Planetary Materials, Papike, J. J. (ed.). *Reviews in Mineralogy*, 36, pp. 1-28.

Celestial voyagers

Bell, J. F., Davis, D. R., Hartman, W. K., Gaffey, M. J. (1989). Asteroids: The Big Picture. In Binzel, R. P., Gehrels, R. and Matthews, M. S. (eds). *Asteroids II*. University of Arizona Press, pp. 921-45.

Brown, P. G. et al. (2000). The fall, recovery, orbit, and composition of the Tagish Lake meteorite: A new type of carbonaceous chondrite. *Science*, 290, pp. 320-25.

Ceplecha, Z. (1961). Multiple fall of Pribram meteorites photographed. *Bulletin of the Astronomical Institute of Czechoslovakia*, 12, pp. 21-47.

Flynn, G. J. (1990). The near-Earth enhancement of asteroidal over cometary dust. *Proceedings of the Lunar and Planetary Science Conference*, 20, pp. 363-71.

Gradie, J. and Tedesco, E. (1982). Compositional structure of the asteroid belt. *Science*, 216, pp. 1405-1407.

Grun, E., Gustafson, B., Mann, I., Baghul, M., Morfill, G. E., Staubach, P., Taylor, A. and Zook, H. A. (1994). Interstellar dust in the heliosphere. *Astronomy and Astrophysics*, 286, pp. 915-24.

Hajdukova, M., Jr. (1994). On the frequency of interstellar meteoroids. *Astronomy and Astrophysics*, 288, pp. 330-34.

Halliday, I., Griffin, A. A. and Blackwell, A. T. (1981). The Innisfree meteorite fall: A photographic analysis of fragmentation, dynamics and luminosity. *Meteoritics*, 16, pp. 153-70.

Halliday, I. et al. (1978). *Journal of the Royal Astronomical Society of Canada*, 72, 15.

McCrosky, R. E., Posen, A., Schwartz, G. and Shao, G.-Y. (1971). Lost City meteorite: its recovery and a comparison with other fireballs. *Journal of Geophysical Research*, 76, pp. 4090-4108.

Oberst, J., Molou, S., Heinlein, D. Gritzner, C., Schindler, M., Spurny, P., Ceplecha, Z., Rendtel, J. and Betlem, H. (1998). The "European Fireball Network": Current status and future prospects. *Meteoritics and Planetary Science*, 33, pp. 49-56.

Taylor, A. D., Baggaley, W. J. and Steel, D. I. (1996). Discovery of interstellar dust entering the Earth's atmosphere. *Nature*, 380, pp. 322-24.

Taylor, S. R. 1994. *Solar System Evolution: A new perspective*. Cambridge University Press, 307 pp.

Vilas, F., Cochran, A. L. and Jarvis, K. S. (2000). Vesta and the vestoids: A new rock group? *Icarus*, 147, pp. 119-28.

Weissman, P. R. (1986). Are cometary nuclei primordial rubble-piles? *Nature*, 320, pp. 242-44.

Wetherill, G. W. (1976). Where do meteorites come from. *Geochimica et Cosmochimica Acta*, 40, pp. 1297-1317.

Wetherill, G. W. and Chapman C. R. (1988). Asteroids and Meteorites. In Kerridge, J. F. and Matthews, M. S. (eds). *Meteorites and the Early Solar System*. University of Arizona Press, pp. 35-67.

Searching for the past

Bevan, A. W. R., Bland, P. A. and Jull, A. J. T. (1998). Meteorite flux on the Nullarbor Region, Australia. In Grady, M. M. et al. (eds). *Meteorites: Flux with Time and Impact Effects*. Geological Society, London, Special Publications, 140, pp. 59-73.

Bischoff, A. and Geiger, T. (1995). Meteorites from the Sahara: Find locations, shock classification, degree of weathering and pairing. *Meteoritics*, 30, pp. 113-22.

Bland, P. A., Sexton, A. S., Jull, A. J. T., Bevan, A. W. R., Berry, F. J., Thornley, D. M., Astin, T. R., Britt, D. T. and Pillinger, C. T. (1998). Climate and rock weathering: a study of terrestrial age dated ordinary chondritic meteorites from hot desert regions. *Geochimica et Cosmochimica Acta*, 62, pp. 3169-84.

Cassidy, W. M., Harvey, R., Schutt, J., Delisle, G. and Yanai, K. (1992). The meteorite collection sites of Antarctica. *Meteoritics*, 27, pp. 490-525.

Grady, M. M. (2000). Catalogue of Meteorites, 5th edn. British Museum Natural History. 689 pp.

Maurette, M. et al. (1991). A collection of diverse micrometeorites from 100 tonnes of Antarctic blue ice. *Nature*, 351, pp. 44-47.

Maurette, M., Jehanno, C., Robin, E. and Hammer, C. (1987). Characteristics and mass distribution of extraterrestrial dust from the Greenland icecap. *Nature*, 328, pp. 699-702.

Nininger, H. H. (1972). *Find a Falling Star*. Paul S. Eriksson

Inc., New York, 254 pp.

Rietmeijer, F. J. M. (1998). Interplanetary dust particles. In Planetary Materials. Papike, J. J. (ed.). *Reviews in Mineralogy*, 36, 2, pp. 2-1–2-95.

Taylor, S., Lever, J. H. and Harvey, R. P. (1998). Accretion rate of cosmic spherules measured at the South Pole. *Nature*, 392, pp. 899-903.

Yoshida, M., Ando, H., Omoto, K., Naruse, R. and Ageta, Y. (1971). Discovery of meteorites near Yamato Mountains, Eastern Antarctica. *Antarctic Record*, No. 39, pp. 62-65.

Yanai, K., Kojima, H. and Haramura, H. (1995). *Catalog of the Antarctic Meteorites*. National Institute of Polar Research, Tokyo, 230 pp.

Weber, D., Zipfel, J. and Bischoff, A. (1999). The Libyan meteorite population. Workshop on Extraterrestrial Materials from Hot and Cold Deserts, Kwa-Maritane, Pilanesburg, South Africa. *Lunar and Planetary Science Contribution*. No. 997. pp. 81-82.

Zolensky, M. E., Wells, G. L. and Rendell, H. M. (1990). The accumulation rate of falls at the Earth's surface: The view from Roosevelt County, New Mexico. *Meteoritics*, 25, pp. 11-17.

Zolensky, M. E., Rendell, H. M., Wilson, I. and Wells, G. L. (1992). The age of the meteorite recovery surfaces of Roosevelt County, New Mexico, USA. *Meteoritics*, 27, pp. 460-62.

Anatomy of a planet

Clayton, R. N. (1993). Oxygen isotopes in meteorites. *Annual Reviews in Earth and Planetary Sciences*, 21, pp. 115-49.

Dodd, R. T. (1986). *Thunderstones and Shooting Stars: The meaning of meteorites*. Harvard University Press, Cambridge, Massachusetts and London, 196 pp.

Hutchison, R. (1983). *The Search for Our Beginning*. British Museum (Natural History)/Oxford University Press, 164 pp.

Patterson, C. C. (1956). Age of meteorites and the Earth. *Geochimica et Cosmochimica Acta*, 10, pp. 230-37.

Ringwood, A. E. (1975). *Composition and Petrology of the Earth's Mantle*. McGraw Hill, New York.

Taylor, S. R. (1988). Planetary compositions. In Kerridge, J. F. and Matthews, M. S. (eds). *Meteorites and the Early Solar System*. University of Arizona Press, Tucson, pp. 512-34.

Taylor, S. R. and McLennan, S. M. (1985). *The Continental Crust: Its Composition and Evolution*, Blackwell, Oxford, England, 312 pp.

Wilde, S. A., Valley, J. W., Peck, W. H. and Graham, C. M. (2001). Evidence from detrital zircons for the existence of continental crust and oceans on Earth 4.4 Gyr ago. *Nature*, 409, pp. 175-78.

Building blocks of planets?

Anders, E. and Grevesse, N. (1989). Abundances of the elements: Meteoritic and solar. *Geochimica et Cosmochimica Acta*, 53, pp. 197-214.

Brearley, A. J. and Jones, R. H. (1998). Chondritic meteorites. In Planetary Materials, Papike, J. J. (ed.). *Reviews in Mineralogy*, 36, 3 pp. 3-1–3-398.

Brownlee, D. E. (1994). The origin and role of dust in the early Solar System. In Zolensky M. E., Wilson, T. L. Rietmeijer, F. J. M. and Flynn, G. J. (eds). Analysis of interplanetary dust, *AIP Conference Proceedings*, American Institute of Physics, 310, pp. 127-43.

Clayton, R. N. (1993). Oxygen isotopes in meteorites. *Annual Reviews of Earth and Planetary Science*, 21, pp. 115-49.

Dodd, R. T. (1981). *Meteorites: A Petrologic-Chemical Synthesis*. Cambridge University Press, 368 pp.

Kalleymeyn, G. W., Rubin, A. E. and Wasson, J. T. (1991). The compositional classification of chondrites: V. The Karoonda (CK) group of carbonaceous chondrites. *Geochimica et Cosmochimica Acta*, 55, pp. 881-92.

Kalleymeyn, G. W., Rubin, A. E. and Wasson, J. T. (1994). The compositional classification of chondrites: VI. The CR carbonaceous chondrite group. *Geochimica et Cosmochimica Acta*, 58, pp. 2873-88.

Kalleymeyn, G. W., Rubin, A. E. and Wasson, J. T. (1996). The compositional classification of chondrites: VII. The R chondrite group. *Geochimica et Cosmochimica Acta*, 60, pp. 2243-56.

H.Y. McSween (1987). Meteorites and their parent planets. Cambridge University Press, 236 pp.

Rietmeijer, F. J. M. (1998). Interplanetary dust particles. In Planetary Materials, Papike, J. J. (ed.). *Reviews in Mineralogy*, 36, 2, pp. 2-1–2-95.

Rubin, A. E. (1997). Mineralogy of meteorite groups. *Meteoritics and Planetary Science*, 32, pp. 231-47.

Rubin, A. E. (1997). Mineralogy of meteorite groups: An update. *Meteoritics and Planetary Science*, 32, pp. 733-34.

Rubin, A. E. (2000). Petrologic, geochemical and experimental constraints on models of chondrule formation. *Earth Science Reviews*, 50, pp. 3-27.

Sorby, H. C. (1877). On the structure and origin of meteorites. *Nature*, 15, pp. 495-98.

Van Schmus, W. R. and Wood, J. A. (1967) A chemical-petrological classification for the chondritic meteorites. *Geochimica et Cosmochimica Acta*, 31, pp. 747-65.

Wasson, J. T. (1985). *Meteorites: Their record of early Solar System history*. W. H. Freeman, New York, 267 pp.

Weisberg, M. K., Prinz, M., Clayton, R. N., Mayeda, T. K., Sugiura, N., Zashu, S. and Ebihara, M. (2001). A new metal-rich chondrite grouplet. *Meteoritics and Planetary Science*, 36, pp. 401-408.

Wood, J. A. (1996). Unresolved issues in the formation of chondrules and chondrites. In R. H. Hewins, R. H. Jones and E. R. D. Scott (eds). *Chondrules and the Protoplanetary Disk*. Cambridge University Press, New York, pp. 55-69.

When asteroids melt

Berkley, J. L., Taylor, G. J., Keil, K., Harlow, G. E. and Prinz, M. (1980). The nature and origin of ureilites. *Geochimica et Cosmochimica Acta*, 44, pp. 1579-97.

Bogard, D. D. and Johnson, P. (1983). Martian gases in an Antarctic meteorite. *Science*, 221, pp. 651-54.

Fogel, R. A., Hess, P. C. and Rutherford, M. J. (1988). The enstatite chondrite-achondrite link. *Lunar and Planetary Science*, 19, pp. 342-43.

Keil, K. (1989). Enstatite meteorites and their parent bodies. *Meteoritics*, 24, pp. 195-208.

McCoy, T. J. and Keil, K., Clayton, R. N., Mayeda, T. K., Bogard, D. D., Garrison, D. H. and Wieler, R. (1997). A petrologic and isotopic study of lodranites: Evidence for early formation as partial melt residues from heterogeneous precursors. *Geochimica et Cosmochimica Acta*, 61, pp. 623-37.

McSween, H. Y., Jr. (1984). SNC meteorites: are they Martian Rocks? *Geology*, 12, pp. 3-6.

McSween, H. Y. Jr and Treiman, A. H. (1998). Martian meteorites. In Planetary Materials, Papike, J. J. (ed.). *Reviews in Mineralogy*, 36, 6, pp. 6-1–6-53.

Mittlefehldt, D. W., Lindstrom, M. M., Bogard, D. D., Garrison, D. H. and Field, S. W. (1996). Acapulco- and Lodran-like achondrites: Petrology, geochemistry, chronology and origin. *Geochimica et Cosmochimica Acta*, 60, pp. 867-82.

Mittlefehldt, D. W., McCoy, T. J., Goodrich, C. A. and Kracher, A. (1998). Non-chondritic meteorites from asteroidal bodies. In Planetary Materials, Papike, J. J. (ed.). *Reviews in Mineralogy*, 36, pp. 4-1 - 4-195.

Nehru, C. E., Prinz, M., Weisberg, M. K., Ebihara, M. E., Clayton, R. N. and Mayeda, T. K. (1992). Brachinites: A new primitive achondrite group. *Meteoritics*, 27, p. 267.

Nehru, C. E., Prinz, M., Weisberg, M. K., Ebihara, M. E., Clayton, R. N. and Mayeda, T. K. (1996). A new brachinite and petrogenesis of the group. *Lunar and Planetary Science*, 27, pp. 943-44.

Okada, A., Keil, K., Taylor, G. J. and Newsom, H. (1988). Igneous history of the aubrite parent asteroid: Evidence from the Norton County enstatite achondrite. *Meteoritics*, 23, p. 59-74.

Righter, K. and Drake, M. J. (1997). A magma ocean on Vesta: Core formation and petrogenesis of eucrites and diogenites. *Meteoritics and Planetary Science*, 32, pp. 929-44.

Scott, E. R. D., Taylor, G. J. and Keil, K. (1993). Origin of ureilite meteorites and implications for planetary accretion. *Geophysical Research Letters*, 20, pp. 415-18.

Spitz, A. H. and Boynton, W. V. (1991). Trace element analysis of ureilites: New constraints on their petrogenesis. *Geochimica et Cosmochimica Acta*, 55, pp. 3417-30.

Taylor, G. J., Keil, K., McCoy, T., Haack, H. and Scott, E. R. D. (1993). Asteroid differentiation: Pyroclastic volcanism to magma oceans. *Meteoritics and Planetary Science*, 28, pp. 34-52.

Takeda, H., Mori, H. and Ogata, H (1989). Mineralogy of augite-bearing ureilites and the origin of their chemical trends. *Meteoritics*, 24, pp. 73-81.

Treiman, A. H. (1989). An alternate hypothesis for the origin of Angra Dos Reis: Porphyry, not cumulate. *Proceedings of the Lunar and Planetary Science Conference*, 19, pp. 443-50.

To the core of the matter

Buchwald, V. F. (1975). *Handbook of Iron Meteorites*. University of California Press, Berkeley, California, 1418 pp.

Burbine, T. H., Meibom, A. and Binzel, R. P. (1996). Mantle

material in the Main Belt: Battered to bits? *Meteoritics and Planetary Science*, 31, pp. 607-20.

Gaffey, M. J. and Gilbert, S. L. (1998). Asteroid 6 Hebe: The probable parent body of the H-type ordinary chondrites and the IIE iron meteorites. *Meteoritics and Planetary Science*, 33, pp. 1281-95.

Goldstein, J. and Ogilvie, R. E. (1965). The growth of the Widmanstätten pattern in metallic meteorites. *Geochimica et Cosmochimica Acta*, 29, pp. 893-920.

Kracher, A., Willis, J. and Wasson, J. T. Chemical classification of iron meteorites: IX. A new group (IIF), revision of IAB and IIICD, and data on 57 additional irons. *Geochimica et Cosmochimica Acta*, 44, pp. 773-87. (and references therein).

Lovering, J. F., Nichiporuk, W. Chodos, A. and Brown H. (1957). The distribution of gallium, germanium, cobalt, chromium, and copper in iron and stony-iron meteorites in relation to nickel content and structure. *Geochimica et Cosmochimica Acta*, 11, pp. 263-78.

McCoy, T. J. (1995). Silicate-bearing IIE irons: Early mixing and differentiation in a core-mantle environment and shock resetting of ages. *Meteoritics*, 30, pp. 542-43.

Mittlefehldt, D. W., McCoy, T. J., Goodrich, C. A. and Kracher, A. (1998). Non-chondritic meteorites from asteroidal bodies. In Planetary Materials, Papike, J. J. (ed.). *Reviews in Mineralogy*, 36, 4, pp. 4-1–4-195.

Narayan, C. and Goldstein, J. I. (1982). A dendritic solidification model to explain Ge-Ni variations in iron meteorite chemical groups. *Geochimica et Cosmochimica Acta*, 46, pp. 259-68.

Rubin, A. E. and Mittlefehldt, D. W. (1992). Evolutionary history of the mesosiderite asteroid: A chronologic and petrologic synthesis. *Icarus*, 101, pp. 201-212.

Rubin. A. E., Jerde, E. A., Zong, P., Wasson, J. T., Wescott, J. W., Mayeda, T. K. and Clayton, R. N. (1986). Properties of the Guin ungrouped iron meteorite: The origin of Guin and of group-IIE irons. *Earth and Planetary Science Letters*, 76, pp. 209-26.

Saikumar, V. and Goldstein, J. I. (1988). An evaluation of the methods to determine the cooling rates of iron meteorites. *Geochimica et Cosmochimica Acta*, 52, pp. 715-26.

Scott, E. R. D., Haack, H. and McCoy, T. J. (1996). Core crystallization and silicate-metal mixing in the parent body of the IVA iron and stony-iron meteorites. *Geochimica et Cosmochimica Acta*, 60, pp. 1615-31.

Scott, E. R. D. (1977). Pallasites: Metal composition, classification and relationships with iron meteorites. *Geochimica et Cosmochimica Acta*, 41, pp. 349-60.

Wilson, L. and Keil, K. (1997). The fate of pyroclasts produced in explosive eruptions on asteroid 4 Vesta. *Meteoritics and Planetary Science*, 32, pp. 813-23.

Wilson, L., Keil, K. and Love, S. J. (1999). The internal structures and densities of asteroids. *Meteoritics and Planetary Science*, 34, pp. 479-83.

Star dust

Alexander, C. M. O'D (1993). Presolar SiC in chondrites: How variable and how many sources? *Geochimica et Cosmochimica Acta*, 57, pp. 2869-88.

Amari, S., Anders, E., Virag, A. and Zinner, E. (1990). Interstellar graphite in meteorites. *Nature*, 345, pp. 238-40.

Amari, S., Lewis, R. S. and Anders, E. (1994). Interstellar grains in meteorites: I. Isolation of SiC, graphite, and diamond; size distributions of SiC and graphite. *Geochimica et Cosmochimica Acta*, 58, pp. 459-70.

Anders, E. and Zinner, E. (1993). Interstellar grains in primitive meteorites: diamond, silicon carbide, and graphite. *Meteoritics*, 28, pp. 490-514.

Black, D. C. and Pepin, R. O. (1969). Trapped neon in meteorites. II. *Earth and Planetary Science Letters*, 6, pp. 395-405.

Burbidge, E. M., Burbidge, G. R., Fowler, W. A. and Hoyle, F. (1957). Synthesis of elements in stars. *Reviews in Modern Physics*, 29, pp. 547-650.

Cameron, A. G. W. (1957). *Stellar evolution, nuclear astrophysics and nucleogenesis.* Chalk River Report, Atomic energy of Canada Limited, CRL-41.

Clayton, R. N., Grossman, L. and Mayeda, T. K. (1973). A component of primitive nuclear composition in carbonaceous meteorites. *Science*, 182, pp. 485-88.

Eberhardt, P. and Geiss, J. (1966). On the mass spectrum of fission xenon in the Pasamonte meteorite. *Earth and Planetary Science Letters*, 1, pp. 99-102.

Grady, M.M. and Wright, I. P. (1995). A cosmic cake mix: Can primitive meteorites tell us what happened before the Solar System was formed? *New Scientist*, 127, No. 1734.

Lewis, R. S., Tang, M., Wacker, J. F., Anders, E. and Steel, E. (1987). Interstellar diamonds in meteorites. *Nature*, 326, pp. 160-62.

Ott, U. (1993). Interstellar grains in meteorites. *Nature*, 364, pp. 25-33.

Reynolds, J. H. (1960). Isotopic composition of primordial xenon. *Physics Review Letters*, 4, pp. 351-54.

Reynolds, J. H. and Turner, G. (1964). Rare gases in the chondrite Renazzo. *Journal of Geophysical Research*, 69, pp. 3263-81.

Russell, S. S., Arden, J. W. and Pillinger, C. T. (1991). Evidence for multiple sources of diamond from primitive chondrites. *Science*, 254, pp. 1188-91.

Russell, S. S., Arden, J. W. and Pillinger, C. T. (1992). The effect of metamorphism on chondritic diamond and silicon carbide (abstract). *Meteoritics*, 27, p. 283.

Sandford, S. A. (1996). The inventory of interstellar materials available for the formation of the Solar System. *Meteoritics and Planetary Science*, 31, pp. 449-76.

Tayler, R. (1989). The birth of the elements. *New Scientist*, 16 December, pp. 19-23.

Zinner, E. (2000). Presolar grains in 2000: Where do we stand? *Meteoritics and Planetary Science*, 35, Supplement A. pp. 177-78.

Zinner, E., Tang, M. and Anders, E. (1989). Interstellar SiC in the Murchison and Murray meteorites: Isotopic composition of Ne, Xe, Si, C. *Geochimica et Cosmochimica Acta*, 53, pp. 3273-90.

Rock of ages

Binns, R. A., Davis, R. J. and Reed, S. J. B. (1969). Ringwoodite, natural $(MgFe)_2SiO_4$ spinel in the Tenham meteorite. *Nature*, 221, pp. 943-44.

Bogard, D. D. (1995). Impact ages of meteorites: A synthesis. *Meteoritics and Planetary Science*, 30, pp. 244-68.

Chen, J. H. and Wasserburg, G. J. (1990). The isotopic composition of Ag in meteorites and the presence of [107]Pd in protoplanets. *Geochimica et Cosmochimica Acta*, 54, pp. 1729-43.

Horan, M. F., Smoliar, M. I. and Walker, R. J. (1998). [182]W and [187]Re-[187]Os systematics of iron meteorites: Chronology for melting, differentiation, and crystallization in asteroids. *Geochimica et Cosmochimica Acta*, 62, pp. 545-54.

Jull, A. J. T., Bevan, A. W. R., Cielaszyk, E. and Donahue, D. J. (1995). [14]C terrestrial ages and weathering of meteorites from the Nullarbor Region, Western Australia. *Lunar and Planetary Institute Technical Report*, 95, 2, pp. 37-38.

Jull, A. J. T., Wlotzka, F., Palme, H. and Donahue, D. J. (1990). Distribution of terrestrial age and petrologic type of meteorites from western Libya. *Geochimica et Cosmochimica Acta*, 54, 2895-98.

Jull, A. J. T., Lal, D., Burr, G. S., Bland, P. A., Bevan, A. W. R. and Beck, W. (2000). Radiocarbon beyond this world. *Radiocarbon*, 42, pp. 151-72.

Lee, T., Papanastassiou, D. and Wasserburg, G. J. (1976). *Geophysical Research Letters*, 3, p. 109.

Lee, T., Papanastassiou, D. and Wasserburg, G. J. (1977). Aluminium-26 in the early Solar System: fossil or fuel? *Astrophysical Journal Letters*, 211, L107-110.

Schmitz, B., Peucker-Ehrenbrink, B., Lindstöm, M. and Tassinari, M. (1997). Accretion rates of meteorites and cosmic dust in early Ordivician. *Science*, 278, pp. 88-90.

Shen, J. J., Papanastassiou, D. A. and Wasserburg, G. J. (1996). Precise Re-Os determinations and systematics of iron meteorites. *Geochimica et Cosmochimica Acta*, 60, pp. 2887-2900.

Smith, J.V. and Mason, B. (1970). Pyroxene garnet transformation in the Coorara meteorite. *Science*, 168, pp. 832-33.

Srinivasan, G., Goswami, J. N. and Bhandari, N. (1999). 26-Al in eucrite Piplia Kalan: Plausible heat source and formation chronology. *Science*, 284, pp. 1348-50.

Urey, H. C. (1955). *Proceedings of the National Academy of Sciences*, 41, p. 127.

Voshage, H. et al. (1983). *Zeitschrift Naturforschung*, 38a.

Decoding the messages

Galy, A., Yound, E. D., Ash, R. D. and O'Nions, K. (2000). The formation of chondrules at high gas pressures in the Solar Nebula. *Science*, 290, pp. 1751-53.

Flynn, G. J. (1996). Sources of 10 micron interplanetary dust: The contribution from the Kuiper belt. In Gustafson, B. Å. S. and Hanner, M. S. (eds). *Physics, Chemistry and Dynamics of Interplanetary Dust*. Astronomical Society Pacific Conference Series, 104, pp. 171-75.

Meibom, A. and Clarke, B. E. (1999). Evidence for the insignificance of ordinary chondritic material in the asteroid belt. *Meteoritics and Planetary Science*, 34, pp. 7-24.

Taylor, S. R. (1994). *Solar System Evolution: A New Perspective*. Cambridge University Press, 307 pp.

Wood, J. A. (1996). Processing of chondritic and planetary material in spiral density waves in the nebula. *Meteoritics and Planetary Science*, 31, pp. 641-45.

Life's rich tapestry

Anders, E. (1991). Organic matter in meteorites and comets: Possible origins. *Space Science Reviews*, 56, pp. 157-66.

Anders, E. and other contributions and responses. (1996). Evaluating the evidence for past life on Mars. *Science*, 274, pp. 2119-25.

Bada, J. L., Galvin, D. P., McDonald, G. D. and Becker, L. (1998). A search for endogenous amino acids in Martian meteorite ALH84001. *Science*, 279, pp. 362-65.

Bailey, J., Chrysostomou, A., Hough, J. H., Gledhill, T. M., McCall, A., Clark, S. Ménard, F. and Tamura, M. (1998). Circular polarization in star formation regions: Implications for biomolecular homochirality. *Science*, 281, pp. 672-74.

Bonner, W. A. (1991). The origin and amplification of biomolecular chirality. *Origins of Life*, 21, pp. 59-111.

Bradley, J. P., Harvey, R. P. and McSween, H. Y. Jr. (1996). Magnetite whiskers and platelets in ALH 84001 martian meteorite: Evidence of vapor phase growth. *Geochimica et Cosmochimica Acta*, 60, pp. 5149-55.

Buick, R., Thornett, J. R., McNaughton, N. J., Smith, J. B., Barley, M. E. and Savage, M. (1995). Records of emergent continental crust 3.5 billion years ago in the Pilbara craton of Australia. *Nature*, 375, pp. 574-77.

Chyba, C. F. and Sagan, C. (1992). Endogeneous production, exogeneous delivery and impact-shock synthesis of organic molecules: an inventory for the origins of life. *Nature*, 355, pp. 125-32.

Cohen, P. (1996). Let there be life. *New Scientist*, 6 July, 1996, pp. 22-27.

Engel, M. H. and Macko, S. A. (1997). Isotopic evidence for extraterrestrial non-racemic amino acids in the Murchison meteorite. *Nature*, 389, pp. 265-67.

Folsome, C. E., Lawless, J. G., Romiez, M. and Ponnamperuma, C. (1973). Heterocyclic compounds recovered from carbonaceous chondrites. *Geochimica et Cosmochimica Acta*, 37, pp. 455-65.

Gladman, B. (1997). Destination: Earth. Martian meteorite delivery. *Icarus*, 130, pp. 228-46.

Goldanskii, V. I. and Kuzmin, V.V. (1991). Chirality and cold origin of life. *Nature*, 352, pp. 114.

Grady, M. M., Wright, I. P. and Pillinger, C. T. (1997). Microfossils from Mars: A question of faith. *Astronomy and Geophysics, The Journal of the Royal Astronomical Society*, 38, No. 1, pp. 26-29.

Groves, D. I., Dunlop, J. S. R. and Buick, R. (1981). An early habitat of life. *Scientific American*, 29, pp. 183-206.

Jull, A. J. T., Courtney, C., Jeffrey, D. A. and Beck, J. W. (1998). Isotopic evidence for a terrestrial source of organic compounds found in Martian meteorites Allan Hills 84001 and Elephant Moraine 79001. *Science*, 279, pp. 366-69.

Jull, A. J. T., Eastoe, C. J. and Cloudt, S. (1997). Isotopic composition of carbonates in the SNC meteorites Allan Hills ALH 84001 and Zagami. *Journal of Geophysical Research*, 102, pp. 1663-69.

Kerridge, J. F. (1993). Origins of organic matter in meteorites. *Proceedings of the National Institute of Polar Research, Symposium on Antarctic Meteorites*, 6, pp. 293-303.

Kvenvolden, K. A., Lawless, J. G., Pering, K., Peterson, E., Flores, J., Ponnamperuma, C., Kaplan, I. R. and Moore, C. B. (1970). Evidence for extraterrestrial amino-acids and hydrocarbons in the Murchison meteorite. *Nature*, 228, pp. 923-26.

Kvenvolden, K. A., Lawless, J. G. and Ponnamperuma, C. (1971). Nonprotein amino acids in the Murchison meteorite. *Proceedings of the National Academy of Sciences*, 68, No. 2, pp. 486-90.

McKay, D. S., Gibson, E. K. Jr, Thomas-Keprta, K. L., Vali, H., Romanck, C. S., Clemett, S. J., Chillier, X., Maechling, C. R. and Zare, R. N. (1996). Search for past life on Mars: Possible relic biogenic activity in martian meteorite ALH84001. *Science*, 273, pp. 924-30.

McNamara, K. J. and Awramik, S. M. (1994). Stromatolites: a key to understanding the early evolution of life. *Science Progress*, 77, pp. 1-20.

Miller, S. L. and Urey, H. C. (1959). Organic compound synthesis on the primitive earth. *Science*, 130, pp. 245-51.

Mojzsis, S. J., Arrhenius, G., McKeegan, K. D., Harrison, T. M., Nutman, A. P. and Friend, C. R. L. (1996). Evidence for life on Earth before 3,800 million years ago. *Nature*, 384, pp. 55-59.

Pering, K. L. and Ponnamperuma, C. (1971). Aromatic hydrocarbons in the Murchison meteorite, *Science*, 173, pp. 237-39.

Ponnamperuma, C. (1972). Organic compounds in the Murchison meteorite. *Annals of the New York Academy of Sciences*, 194, pp. 56-70.

Schopf, J. A. (1993). Microfossils of the Early Archean Apex Chert: New evidence of the antiquity of life. *Science*, 260, pp. 640-46.

Siering, P. L. (1998). The double helix meets the crystal lattice: The power and pitfalls of nucleic acid approaches for biomineralogical investigations. *American Mineralogist*, 83, pp. 1593-1607.

Treiman, A. H. (1999). Microbes in a martian meteorite? *Sky and Telescope*, 97, No. 4, pp. 52-58.

Uwins, P. J. R., Webb, R. I. and Taylor, A. P. (1998). Novel nano-organisms from Australian sandstones. *American Mineralogist*, 83, pp. 1541-50.

Urey, H. C. and Lewis, J. (1966). Organic matter in carbonaceous chondrites. *Science*, 152, pp. 102-104.

Blasts from the past

Alvarez, L. W., Alvarez, W., Asaro, F. and Michel, H. (1980). Extraterrestrial cause for the Cretaceous/Tertiary extinction, *Science*, 208, pp. 1095-1108.

Bohor, B. F., Modreski, P. J. and Foord, E. E. (1987). Shocked quartz in the Cretaceous-Tertiary boundary clays: Evidence for a global distribution. *Science*, 236, pp. 705-709.

Bourgeois, J., Hansen, T. A., Wiburg, P. L. and Kauffman, E. G. (1988). A tsunami deposit at the Cretaceous-Tertiary boundary in Texas. *Science*, 241, pp. 567-70.

Chapman, C. R. and Morrison, D. (1994). Impacts on Earth by asteroids and comets: assessing the hazard. *Nature*, 367, pp. 33-34.

Frankel, C. (1999). *The End of the Dinosaurs: Chicxulub Crater and Mass Extinctions*. Cambridge University Press, 223 pp.

Grieve, R. A. F. (1998). Extraterrestrial impacts on Earth: the evidence and the consequences. In Grady, M. M. et al. (eds). *Meteorites: Flux with Time and Impact Effects.* Geological Society, London, Special Publications, 140, pp. 105-31.

Hildebrand, A. R., Pilkington, M., Ortiz-Aleman, C., Chavez, R. E., Urrutia-Fucugauchi, J., Connors, M., Graniel-Castro, E., Camara-Zi, A., Halpenny, J. F. and Niehaus, D. (1998). Mapping Chicxulub crater structure with gravity and seismic reflection data. In Grady, M. M. et al. (eds). *Meteorites: Flux with Time and Impact Effects.* Geological Society, London, Special Publications, 140, pp. 155-76.

Hildebrand, A. R., Penfield, G. T., Kring, D. A., Pilkington, M., Camargo, Z. A., Jacobsen, S. and Boynton, W. (1991). Chicxulub crater: a possible Cretaceous-Tertiary boundary impact crater on the Yucatan Peninsula, Mexico. *Geology*, 19, pp. 867-71.

Isett, G. A. (1990). The Cretaceous/Tertiary boundary interval, Raton Basin, Colorado and New Mexico, and its content of shock-metamorphosed minerals: Evidence relevant to the K-T boundary impact extinction theory. *Geological Society of America Special Paper*, 249.

Koeberl, C., Sharpton, V. L., Murali, A. V. and Burke, K. (1990). The Kara and Ust-Kara impact structures (USSR) and their relevance to the K/T boundary. *Geology*, 18, pp. 50-63.

McNamara, K. and Bevan, A. (1991). *Tektites.* Western Australian Museum Publication, 28 pp.

Sharpton, V. L., Dalrymple, G. B., Marin, L. E., Ryder, G., Schuraytz, B. C. and Urrutia-Fucugauchi, J. (1992). New links between the Chicxulub impact structure and the Cretaceous-Tertiary boundary. *Nature*, 359, pp. 819-21.

Shoemaker, E. M., Wolfe, R. F. and Shoemaker, C. S. (1990). Asteroid and comet flux in the neighbourhood of Earth. In Sharpton, V. L. and Ward, P. D. (eds). *Global Catastrophes in Earth History; An Interdisciplinary Conference on Impacts, Volcanism and Mass Mortality.* Geological Society of America Special Paper, 247, pp. 155-70.

Sigurdsson, H., D'Hondt, S., Arthur, M. A. et al. (1991). Glass from the Cretaceous-Tertiary boundary in Haiti. *Nature*, 349, pp. 482-87.

Wallace, M. W., Gostin, V. A. and Keays, R. R. (1996). Sedimentology of the Neoproterozoic Acraman impact-ejecta horizon, South Australia. *AGSO Journal of Australian Geology and Geophysics*, 16, pp. 443-51.

Williams, G. E. (1994). Acraman, South Australia: Australia's largest meteorite impact structure. *Proceedings of the Royal Society of Victoria*, 106, pp. 105-27.

To the future

Bland, P. A., Bevan A. W. R. and Jull, A. J. T. (2000). Ancient meteorite finds and the Earth's surface environment. *Quaternary Research*, 53, pp. 131-42.

Bland, P. A. and Smith, T. B. (2000). Meteorite accumulations on Mars. *Icarus*, 144, pp. 21-26.

Bland, P. A., Bevan, A. W. R. and Verveer, A. (1999). A Nullarbor Camera Network. *Meteoritics and Planetary Science*, 34, A12.

Zolensky, M. E., Pieters, C., Clark, B. and Papike, J. J. (2000). Small is beautiful: The analysis of nanogram-sized astromaterials. *Meteoritics and Planetary Science*, 35, pp. 9-29.

WEBSITES AND LINKS

ESA future missions	http://sci.esa.int/home/futuremissions/
Life from Mars?	http://www.jsc.nasa.gov/pao/flash/marslife/faq.htm
Lunar and Planetary Institute	http://cass.jsc.nasa.gov/lpi.html
Mars Express and Beagle2	http://www.beagle2.com
Martian meteorites	http://www.jpl.nasa.gov/snc/
Meteoritical Society	http://www.uark.edu/campus-resources/metsoc/index1.htm
NASA	http://www.nasa.gov/NASA_homepage.html/
Stardust Mission	http://spacelink.msfc.nasa.gov/NASA.Projects/Space.Science/Solar.System/Stardust.Mission/

ablation Stripping of material from the surface of an object, as when (a) friction with the atmosphere causes the surface of a *meteorite* to melt and material is removed, (b) fierce Antarctic winds erode and strip the ice cap.

achondrite A *stony meteorite* (lacking *chondrules*) made of the products of melting.

agate Common name for a banded variety of rock made of tiny crystals of quartz.

angular momentum The quantity of motion in a body measured by multiplying its mass by its angular velocity.

Apollo asteroids Group of small *asteroids* in orbits crossing that of Earth.

asteroids Small planet-like bodies (or large fragments thereof) that orbit the Sun, mostly occupying the region between Mars and Jupiter.

ataxites A group of *iron meteorites* containing more than about 13 per cent nickel by weight, without any internal structure visible to the naked eye.

atom Smallest unit of an *element* that can take part in chemical change.

basalt (adj basaltic) Dark volcanic rock made mainly of the *minerals olivine, pyroxene* and *feldspar*.

binary star Two stars locked in *orbit* around each other.

blue ice Compressed ice found in ice sheets.

bolide Atmospheric *fireball* accompanied by loud *sonic booms*.

breccia Rock made up of angular fragments derived from pre-existing rocks.

chondrites The *stony meteorites*, usually characterised by *chondrules*. Chondrites are primitive aggregates of early *Solar System* materials not subsequently melted.

chondrules Small particles, usually smaller than 1 mm and often near-spherical. They can make up to 80 per cent of the volume of some *meteorites*. Chondrules crystallised from once molten, or partly molten droplets.

coma That glowing part of a *comet* surrounding the nucleus, caused by radiation from the Sun.

comets Bodies formed of ice and rock that come from the extreme regions of the *Solar System*. Comets become visible when their *orbits* change bringing them into the inner Solar System where heating from the Sun causes their ices to vaporise forming a *coma*, and releasing gas and dust into tails. Cometary tails always point away from the Sun.

compound A substance resolvable into two or more *elements* united such that the whole has properties of its own that are not necessarily those of its components.

cosmic rays Energetic particles of matter travelling close to the speed of light. Their source is unknown, although they probably originate within our galaxy.

cosmic ray exposure age The length of time during which a small body in space is bombarded by *cosmic rays*. The interaction of cosmic rays with solid matter produces nuclear reactions to depths of tens of centimetres. The amount of secondary radiation so produced is proportional to the length of time of exposure, so the 'age' of ejection of a metre-sized rock from a larger parent body can be calculated.

cosmic spherules Small, rounded bodies, less than 1 mm across, produced by melting of *micrometeoroids* in the atmosphere.

Cretaceous Period of geological time between 135 and 65 million years ago.

crystal Naturally occurring body with plane surfaces that are an outward expression of a regular interior arrangement of *atoms*. Crystallisation is the process by which crystals form.

crystalline The property of a substance with a regular, ordered, interior arrangement of *atoms* (cf *glass*).

cumulate *Igneous* rock formed by the separation and accumulation of *minerals* crystallising from molten magma.

decay rate The rate at which radioactive *atoms* of certain *isotopes* of *elements* decay to isotopes of the same, or different elements.

differentiation The process of developing more than one rock type from a common magma.

diogenite *Igneous achondrite* made mainly of an iron- and magnesium-bearing *silicates* belonging to the *pyroxene* group of minerals.

electron Negatively charged particle inside *atoms* (ie subatomic).

electrophonic sounds 'Hissing' noises associated with the falls of *meteorites*. Their origin is disputed but may be related to electromagnetic radiation generated by *fireballs*.

element A rudimentary substance that cannot be resolved by chemical means into simpler substances.

escape velocity The velocity a body has to reach in order to escape the gravitational pull of another, usually planetary or stellar, body. The Earth's escape velocity is 11.2 km/s.

eucrite *Igneous achondrite* of *basaltic* composition.

explosion crater A crater caused by the collision of a large natural body from space impacting at high velocity with the Earth's surface.

extra-terrestrial Originating from outside of the Earth.

fall A *meteorite* recovered quickly after its fall is witnessed (cf *find*).

feldspars A mineral group of *silicate minerals* commonly containing calcium, sodium, potassium and aluminium.

find A *meteorite* recovered by chance discovery without observing its *fall*.

fireball The light phenomena in the atmosphere caused by frictional heating of an infalling *meteoroid*.

fractionation The separation, by a variety of means, of a quantity of material that is chemically distinct from a larger quantity of parental material (cf *differentiation*).

fusion crust A coating of solidified surface melt on a *meteorite* formed by friction during its passage through the Earth's atmosphere.

galaxy A large conglomeration of stars, gas and dust in the Universe.

glass A non-*crystalline* solid formed by the freezing of molten material.

half-life The time taken for half the number of radioactive *atoms* of an element to decay.

hexahedrite An *iron* containing 5 per cent nickel by weight, made principally of single *crystals* of kamacite (low nickel metal).

howardite An *achondrite* made of fragments of other types, commonly *eucrite*, *diogenite*, *mesosiderite* and *chondrite*. Howardites are broken meteoritic soils from the surface of an *asteroid*.

igneous rocks Rocks produced directly or indirectly from the solidification of molten magma.

inclusions Any bodies of material enclosed by a different body of material.

individual A single *meteorite* from a shower.

interplanetary dust *Micrometeorites* little altered by passage through the atmosphere.

interstellar dust Small particles of dust originating from beyond the *Solar System*. Currently, their existence in the Solar System is inferred from the velocities of some particles that exceed the *escape velocity* of the Sun.

interstellar medium Region between the stars consisting of gas and dust, together with magnetic fields and cosmic rays.

iron meteorites (or irons) *Meteorites* made mainly of iron-nickel metal.

iron-shale The product of complete alteration of an *iron meteorite* to iron oxides by weathering on Earth.

ironstone Common name for concretions rich in iron oxide that occur naturally in a variety of rocks on Earth.

isotope One of two or more species of the same chemical *element*. Isotopes of elements have the same number of *protons* in the nucleus, but different numbers of *neutrons* and hence atomic weights. Although isotopes of the same element have the same chemical properties, they can be separated by their physical properties, usually by mass.

loëss Wind-borne sediment.

magma Molten rock.

mesosiderite A group of *stony-iron meteorites*.

metallurgy The study of natural and man-made metals and alloys.

metamorphic (of rocks) Produced from processes of change (metamorphism), usually by heat and/or pressure.

meteor Streak of light in the Earth's upper atmosphere produced by the entry of a small body from space.

meteorite (adj meteoritic) A natural body from space that survives its passage through the atmosphere to land on the Earth's surface.

meteoriticist One who studies *meteorites*.

meteoritics The study of *meteorites* and associated phenomena.

meteoroid A small natural object in space.

micrometeorites Small (less than 1 mm) particles of dust that pass through the Earth's atmosphere largely unaltered.

Milky Way The *galaxy* of stars to which our *Solar System* belongs.

minerals Naturally occurring substances with a definite chemical composition and a regular, *crystalline* internal arrangement of *atoms*.

mineralogy The study of *minerals*.

molecules The smallest particles of any substance that retain the properties of that substance, made of *atoms* in specific combinations of *elements*.

nebula A body of gas and dust in the Universe.

neutron Sub-atomic particle with no electrical charge.

nova (pl novae) A star that brightens over a short time (days) to thousands of times their normal brightness. The decline in brightness may take many decades. Novae are less spectacular than *supernovae*, and eject only a small part of their mass into space.

nuclide A species of *atom* characterised by the number of *protons* and *neutrons* in its nucleus.

octahedrite *Iron meteorite* displaying a visible *Widmanstätten pattern*.

olivine Group of mainly iron- and magnesium-bearing *silicate* minerals with the general composition $(Fe,Mg,Mn)_2SiO_4$.

Oort cloud Postulated region surrounding the *Solar System* occupied by comets.

orbit The regular path described by a body around its gravitational focus; as of the Earth around the Sun, or the Moon around the Earth.

orientated meteorite A *meteorite* that has a conical shape resulting from uniform *ablation* in the atmosphere.

oxidation (-ised) The process by which *atoms* of an *element* lose *electrons*. Oxidation and its opposite, *reduction*, essentially operate in tandem. When metallic iron rusts, it loses electrons to oxygen and water. In so doing, oxygen and water gain electrons and are effectively 'reduced' as the iron combines with them to produce iron oxides. In a reverse of the process, iron oxide can be reduced to metallic iron by contact with hot hydrogen gas. A by product of this reaction is water.

PAH Polycyclic aromatic hydrocarbons.

pallasites *Stony-iron meteorites* made mainly of iron-nickel metal, *olivine* and *pyroxene* minerals.

peptide bond Amino acids have amino (NH_2) and carboxyl (COOH) groups. Two amino acid *molecules* can be bonded (joined) by the removal of water (H_2O): H from the amino and OH from carboxyl groups. The resulting construction -CO-NH- is called the peptide bond. A series of such bonds can produce protein-like chains.

periodic comet A *comet* in a regular *orbit* around the Sun that brings it periodically into the inner *Solar System*. Halley's Comet has an orbit with a period (one complete orbit) of 76 years.

planetesimals (also *planetoid* or *asteroid*) Small, cold, planet-like bodies that formed by *accretion* in the early *Solar System*.

protons Sub-atomic particles with a positive electric charge.

pumice Light-coloured, frothy volcanic rock made of glass.

pyrite Brassy yellow iron disulfide mineral (FeS_2).

pyroclastic Made up of *igneous* fragments and dust erupted from volcanoes.

pyroxene A group of rock-forming *silicates* commonly containing iron, magnesium, calcium, aluminium and sodium.

radiogenic Product of a radioactive process (eg radiogenic lead).

radionuclide A radioactive *nuclide*.

reduction (reduced) The opposite chemical reaction to *oxidation*.

refractory (of elements) With high boiling and melting temperatures. Also said of a substance from which it is difficult to extract its constituents.

regmaglypts Thumbprint-like depressions on a *meteorite* caused by the uneven flow of air during surface melting and *ablation* in the atmosphere.

retrograde orbit An *orbit* of an object in the *Solar System* that is in the opposite sense to that of the planets.

shatter-cones Cone-like, striated fractures in rocks subjected to high shock pressures.

shooting stars Common name for the light phenomena (*meteors*) caused by the high-speed passage of small particles in the upper atmosphere.

siderophile (of elements) With an affinity, generally weak, to oxygen and sulfur, but that is readily soluble in molten iron (eg nickel, cobalt, platinum metals, gold and tin).

silicates Group containing a large number of common rock-forming *minerals*, the *crystalline* structures of which are based on repetition of the SiO_4 arrangement.

SNC *Achondrite meteorites* from Mars (acronym from the Shergotty, Nakhla and Chassigny meteorites).

solar nebula The postulated cloud of gas and dust from which the *Solar System* formed.

solar wind Radiating atomic particles (mainly protons) streaming from the Sun.

sonic boom The loud noise associated with an object breaking the sound barrier in the atmosphere.

stone See *stony meteorites*.

stony meteorites (also stones) *Meteorites* made predominantly, but not exclusively, of *silicate* minerals.

stony-iron meteorites (also stony-irons) Meteorites made of iron-nickel metal and *silicate* minerals in roughly equal proportions.

strewn field The usually elliptical area over which fragments from a shower of *meteorites* are distributed on the ground. Also applied to the ground distribution of *tektites*.

stromatolites Domed sedimentary structures built by the activity of micro-organisms.

supernova (pl supernovae) An exploding star observed astronomically as a bright light. Supernovae manufacture and distribute heavy *elements*, ejecting a large portion of their mass into space.

tektite Naturally occurring, once airborne, *glass* ejected by explosive *meteorite* impact.

terrestrial age The length of time a *meteorite* has resided on Earth.

vapour deposition The formation of solid material from a vapour.

Widmanstätten pattern Regular trellis work of interlocking plates of iron-nickel *minerals* formed with extremely slow cooling from high temperatures. It can be observed by acid etching of metal-rich *meteorites*.

Zodiacal Light Glow of light variably observed from Earth at dusk and dawn caused by the scattering of the Sun's light by dust in the inner *Solar System*.

ACKNOWLEDGMENTS

This book rests on a great volume of published research. At an early stage in writing, we realised that individual acknowledgments of the many scientists involved would make the text virtually unreadable, the science lost amongst the congratulation. Mindful of the need to acknowledge those who did the work, made the breakthrough, or documented the discovery, we have provided a list of references that acts both as an acknowledgment to those concerned, and as a guide for those interested to pursue the subject.

Many people contributed directly or indirectly to the content of this book and these are acknowledged individually. We thank all those who gave freely of their knowledge, expertise and images that enrich the text. We are particularly thankful to the following (in alphabetical order): Professor Stanley Awramik, Dr Hans Betlem, Mrs Jennifer Bevan, Dr Phil Bland, Dr John Bradley, Dr Andrew Buchanan, Dr Vagn Buchwald, Dr Li Chunlai, the late Mr Bill Cleverly, Mr Brian Clifton, Mr Mike Freeman, Mr Hossein Gahrib, Mr Robert Garvey, Dr Monica Grady, Mr Dieter Heinlein, Professor Paul Hodge, Dr Robert Hough, Dr Russell Hudson, Dr Robert Hutchison, Dr Glenn Izett, Dr Timothy Jull, Dr H. U. Keller, Dr Christian Koeberl, Dr Candace Kohl, Dr Leo Laden, Mr Vic Levis, Professor Michael Lipschutz, Dr Joe McCall, Dr Ken McNamara, Dr Robert McNaught, Dr David Malin, Dr Ursula Marvin, Dr Michel Maurette, Professor Colin Pillinger, Dr Judith Pillinger, the late Mr Keith Quartermaine, Dr Giles Graham, Mr Gary Jones, Ms Sheena Elliott, Mr Peter Downes, Mr Peter Holst, Dr Jochen Schlütter, Dr Birger Schmitz, Mr Robert Shaw, Dr Carolyn Shoemaker, the late Dr Eugene Shoemaker, Mr Tom Smith, Dr Mario Tassinari, Mr H. M. Walker, Professor Simon Wilde, Dr Jeff Wynn, Dr Keizo Yanai, Dr Tony Yeates, Professor Ouyang Ziyuan, and Dr Mike Zolensky.

Those individuals and organisations contributing images are acknowledged individually. Special thanks go to Mrs Danielle West for preparing the illustrations and providing logistical help. Additional photographs were provided by Mr Geoff Deacon, Ms Kristine Brimmell and Mr Douglas Elford (Western Australian Museum).

We are particularly indebted to those who critically read and improved earlier versions of the manuscript, including: Emeritus Professor Ross Taylor, Dr Ken McNamara, Dr Victor Gostin, Dr Malcolm Walter, Dr Robert Hough, Mrs Jennifer Bevan, Dr Mark Sephton and two anonymous reviewers.

ILLUSTRATION SOURCES

Ancient beliefs

Page 12   Right. After Bjorkman, 1973.
Page 12-13   Centre. From Wainwright, 1912.
Page 13   Left. Courtesy of Meteoritics and Planetary Science.
Page 14   Top. Courtesy of Meteoritics and Planetary Science, after Dietz and McHone, 1974.
Page 14   Bottom. Courtesy of J. Wynn.
Page 15   Courtesy of P. Hodge.
Page 16   Top. From Ward, 1904.
Page 16   Bottom. From Ross, 1819.
Page 17   Top left. Courtesy of L. Laden.
Page 17   Top right. Courtesy of the Natural History Museum, London.
Page 17   Bottom. Courtesy of V. Buchwald.
Page 19   Right. From the annals of Semen Remezov.
Page 20   Courtesy of the Deutschen Staatsbibliothek, Berlin.
Page 22   Top. Courtesy of C. T. and J. Pillinger.
Page 23   Main picture. Courtesy of C. Kohl.

Rocks from space

Page 26   Photograph reproduced with permission of the Trustees of the Museum and Galleries of Northern Ireland and © H. M. Walker.
Page 27   Courtesy of B. Clifton.
Page 28   Bottom left. Courtesy of H. Betlem, © H. Betlem.
Page 28   Bottom right (two photographs). Courtesy of H. A. Gharib, © H. A. Gharib.
Page 29   Top. Courtesy of A. T. Kearsley, Oxford Brookes University, and G. A. Graham, Open University, UK.
Page 29   Bottom. Courtesy of J. P. Bradley .
Page 30   Bottom. After Halliday and Griffin, 1982 and Heide and Wlotzka 1995.
Page 32   © R. Garvey, reproduced with permission.
Page 35   Top. Courtesy of C. Kohl.
Page 35   Far left. Courtesy of P. Holst.
Page 35   Bottom. D. Elford and K. Brimmell.
Page 36   Courtesy of Ouyang Ziyuan and Li Chunlai, Institute of Geochemistry, Academia Sinica.
Page 37   G. Deacon
Page 38   Courtesy of J. Pitt and A. Pellegrini.
Page 39   Courtesy of Kevron Aerial Services, Perth.

Celestial voyagers

Page 43   © NASA photo.
Page 44   © Anglo Australian Observatory, photo by D. Malin.
Page 46   Courtesy of D. Heinlein, DLR, European Network of Fireball Photography and Meteorite Recovery.
Page 47   Bottom. Courtesy of the Smithsonian Institution, National Museum of Natural History.
Page 48   Left. © European Space Agency/Max Planck Institut fur Aeronomie, courtesy of H. U. Keller.
Page 48–49   © R. H. McNaught, reproduced with permission.
Page 50   Courtesy of Perth Observatory.
Page 51   Insert. After Weissman, 1986.

Searching for the past

Page 56   Courtesy of P. A. Bland.
Page 57   Middle and bottom. Courtesy of K. Yanai.
Page 60   Bottom. Courtesy of S. Elliot.
Page 61   Top. Courtesy of J. Schlütter.
Page 62   Top. Courtesy of M. Maurette.
Page 62   Bottom. Courtesy of T. Smith.
Page 63   Top. Courtesy of M. Maurette.
Page 63   Bottom. © NASA photo.

Anatomy of a planet

Page 67   © NASA photo
Page 70 and 72 G. Deacon
Page 77   Main picture. © NASA photo.
Page 77   Inset bottom. After McSween and Treiman, 1998.
Page 78   © NASA photo.
Page 79   © NASA photo.

Building blocks of planets?

Page 82   Bottom. Courtesy of R. Hudson.
Page 82   Top. K. Brimmell.
Page 83   Courtesy of R. Hudson.
Page 85   Top. Courtesy of R. Hudson.
Page 85   Middle. K. Brimmell.
Page 86   Courtesy of R. Hudson.

INDEXES